VOICES
IN THE OCEAN

ALSO BY SUSAN CASEY

The Devil's Teeth: A True Story of Obsession
and Survival Among America's Great White Sharks

The Wave: In Pursuit of the Rogues, Freaks,
and Giants of the Ocean

VOICES
IN THE OCEAN

≋

A Journey into the
Wild and Haunting World
of Dolphins

≋

SUSAN CASEY

Doubleday

New York London Toronto Sydney Auckland

Chapter-opener photograph of wild spinner dolphins by James R.D. Scott /
    Moment Open / Getty Images
Jacket design by John Fontana
Jacket photograph © Willyam Bradberry / Shutterstock

Page 303 constitutes an extension of this copyright page.

Library of Congress Cataloging-in-Publication Data
Casey, Susan.
Voices in the ocean : a journey into the wild and haunting world of dolphins /
    Susan Casey.
    pages cm
Includes bibliographical references.
ISBN 978-0-385-53730-8 (hardcover)—ISBN 978-0-385-53731-5 (eBook)
1. Dolphins. I. Title.
QL737.C432C34 2015
599.53—dc23   2015011763

MANUFACTURED IN THE UNITED STATES OF AMERICA

10 9 8 7 6 5 4 3 2 1

First Edition

For Rennio, always

The world is full of magic things, patiently waiting for our senses to grow sharper.

W. B. YEATS

CONTENTS

A NOTE ON DOLPHIN SPECIES

All of the species I write about in these pages are toothed whales (Odontoceti), one of two branches of cetaceans, a marine mammal group that includes whales, dolphins, and porpoises. The word "cetacean" comes from the Latin *cetus*, which means "whale," and the Greek *ketos*, with its rather less flattering translation: "sea monster."

Delphinidae, or oceanic dolphins, are the largest family of toothed whales, containing approximately thirty-seven species that range from the four-foot-long Hector's dolphin to the twelve-foot bottlenose dolphin to the twenty-five-foot orca, or killer whale. The Delphinidae also include pilot whales, melon-headed whales, false killer whales, and pygmy killer whales. Among this group, the word "whale" indicates a creature's size (very big, or at least bigger than your average dolphin), rather than being a precise scientific description.

However, not all dolphins belong to the Delphinidae. There are five species of river dolphins—remarkable, prehistoric-looking animals like the Amazon boto, the Ganges River dolphin, and the now-extinct baiji, formerly found in China's Yangtze River.

I have also included beluga whales in this narrative. These white whales are one of two family members of Monodontidae, along with narwhals.

Porpoises are a separate band entirely, although in the past the words "dolphin" and "porpoise" were often used interchangeably. The seven species of porpoises, or Phocoenidae, are smaller, and distinct from dolphins.

VOICES
IN THE OCEAN

HONOLUA

The road to Honolua Bay was red dirt against a gray sky, and it wound up the bluff in a series of steep switchbacks. I pulled over at the top, where the grade leveled off in a clearing. Usually this lookout was packed with cars, trucks, surfers scouting the break, tourists taking photos, but on this storm-tossed day no one else was around. I got out and walked to the edge of the embankment. Below me, small waves broke on jagged lava rocks, their crests whipped white by the wind.

Low clouds pressed down, turning the bay—a crescent that usually glimmered in a spectrum of blues, from pale aqua to inky cobalt—a dull slate color. Even on brighter days, Honolua was a heavy spot. In centuries past the bay was a place of worship for Hawaiians, who pushed off from its lee in their voyaging canoes and made offerings to their gods in stone-heaped *heiaus* above its shoreline. There were no wide sandy beaches here, just a jigsaw of rocks tumbling down to the water, disappearing beneath the surface where they formed a reef, shallow at first, then dropping off to a darker realm.

Conditions were crummy, but I had driven a long way to get here and this particular bay was known for its marine beauty, its profusion of corals and creatures, so I didn't want to leave without at least getting wet. I wouldn't have another chance for a long while: by tomorrow at this time I would be flying back to New York City. I was aware that a recent flurry of shark attacks had people thinking twice about going into the water alone, or even at all. Suddenly, it seemed, everyone on Maui had realized they shared the ocean with large, occasionally snappish beasts. Around the island there was a deep need for a clear explanation—Too many sea turtles? Not enough fish? Climate change? Were the planet's poles, perhaps, flipping?—some way to figure out the situation, wrap it up, get back to a time when sharks didn't occupy the headlines every other day. I stood there in the wind considering my options, and after a few moments spent listening to my mind spin tales of lost limbs, sheared arteries, nothing left of me but a few scraps of bathing suit, I picked my way down the path and across the rocks, stepped into the water, and began to swim across the bay.

In the water I would only be as alone as I felt on land, anyway. If something happened to me out here, I wasn't sure I cared.

I had lived with that feeling, a dull indifference to pretty much everything, for almost two years, since my father had died of a heart attack. He was seventy-one years old and athletic and strong, and when his heart's electrical system seized up, he had been at our family's summer cottage, walking down to the dock to take his seaplane out for a spin. The doctors said it probably took him five seconds to die.

In some dim, distant corner of my mind I had always known, as every person does, that my father wouldn't be around forever, but the idea of losing him was so huge and overwhelming that I never gave it any space. It lived inside my head as the most horrifying thing I could possibly imagine, the monster I hoped never to face. By the time I hit my forties—divorced, childless by choice, restless by nature—I knew that my father was the central figure in my life, the rock, the anchor, the wise man whose presence allowed me to roam the world making mistakes and having adventures because I could always trust he'd be there at the end to help me make sense of it all. I could no more imagine a life without him than I could imagine life without my torso. And yet, here I was.

But a strange thing happens when your worst nightmare is realized: nothing much is left to scare you. After the initial tsunami of grief, I found myself walking calmly into situations that would have previously terrified me: a solo swim, at dusk, in prime tiger shark territory, for instance. Fear was replaced by an ever-present numbness.

As I headed across the mouth of the bay I veered slightly south, out to sea, until I was a half mile offshore. Treading water, I cleared my goggles and looked around. I could faintly see the bottom, unperturbed and sandy, and conditions were smoother out here, so I didn't turn back. I kept swimming. Some people crave illicit substances when upset; my drug of choice is saltwater. The ocean's vast blue country was either peace or oblivion, I wasn't sure which, but both of those possibilities worked for me.

I was about to head back when a movement caught my eye: a large, shadowy body passed diagonally below me. Then, a jutting dorsal fin; beside it something white flashed. Streaks of sunlight had filtered through the clouds and suddenly the water was illuminated. My adrenaline surged as the creatures revealed themselves.

It was a pod of spinner dolphins, forty or fifty animals, swimming toward me. They materialized from the ocean like ghosts, shimmering in the ether. One moment they were hazily visible, then they were gone, then they reappeared on all sides, surrounding me. I had never been this close to dolphins before, and I was amazed by their appearance. One of the bigger spinners approached slowly, watching me. For a moment we hung there in the water and looked at one another, exchanging what I can only describe as a profound, cross-species greeting. His eyes were banded subtly with black, markings that trailed to his pectoral fins like an especially delicate bank robber's mask. I wondered if he was the pod's guardian, if the others followed his lead. The dolphins were traveling in small but distinct clusters—couples, threesomes, klatches of four or five—and within those little groups they maintained close body contact. I saw fins touching like handholding, bellies brushing across backs, heads tilted toward other heads, beaks slipped under flukes.

The entire group could have darted away in an instant, but they chose instead to stay with me. Spinners are known for their athletics, rocketing out of the water in aerial leaps whenever the urge strikes, but these dolphins were relaxed. They showed no fear, despite the presence of several baby spinners tucked in beside their mothers, replicas the size of bowling pins. The dolphins had simply enfolded me in their gathering, and I could hear their clicks and buzzes underwater, their cryptic aquatic conversation.

I dove ten feet down and the big dolphin appeared beside me again, even closer. He had coloration like a penguin's, dark on top and tuxedo white on his belly, with a long, slender beak. At eight feet long he was a powerful animal, but nothing in his body language suggested hostility. We stayed together for maybe ten minutes but the meeting felt eternal, as though time were suspended in the water with us. The ocean rose and fell rhythmically, almost hypnotically, but I had no point of reference, no horizon. There was no land, no sky. Everything glowed, as if viewed through a lush blue prism. The dolphins watched me watching them. They moved with an unearthly grace, as though they were more presence than form. I swam with the spinners until they headed into deeper waters, where the light fell off to nowhere in long, slanting rays. The last thing I saw before they vanished back into their world was their tails, moving in unison.

≈

After my encounter with them, I thought of the dolphins often. Not just for hours or days afterward, but for weeks and months. I thought of them at night as I was going to sleep—remembering their languid swimming motions made me relaxed and drowsy and calm. I thought of the dolphins after I left Hawaii and returned to Manhattan, where life was anything but relaxed and drowsy and calm, and where the luminous blues of the Pacific Ocean were a distant memory. In my thirty-sixth-floor office, in a towering glass and steel building in the city's midtown, I thumbtacked pictures of dolphins to the wall behind my desk so I could look at them while I made phone calls.

However brief my dolphin visitation had been, it was stuck to me, lodged inside my head. It was as though I'd been hit by lightning and that one strike had zapped clean through my brain, replacing its usual patterns and wavelengths and nerve impulses with a dolphin highlight reel. I couldn't forget the way the pod had sized me up, or their peculiar squeaking, creaking language, or how ridiculously *fun* it was to just cruise along with them. I got the impression there was somebody home behind each set of eyes, and the effect was surreal. I'd met other intriguing sea creatures, some shy and some lordly, some beautiful and some that only Mother Nature could love, but none of them had the same presence as the dolphins—not the Buddha-faced puffer fish with its wise eyes and tiny, whirring fins, or the spotted eagle ray that resembled an alien spacecraft, or the bullheaded ulua, a muscular game fish you wouldn't want to meet in a dark alley. Next to the fluid, social dolphins, the great white sharks I'd seen looked so metallic I thought they must have rivets. They were undersea Hindenbergs, majestic but not heartwarming, and if you had a personal encounter with one of them it was unlikely to be a calming experience.

At the risk of falling down the rabbit hole—a place you can easily go with dolphins, I would soon learn—my most enduring impression was how *otherworldly* the animals were. As they swam by me, they seemed to exist in a more hazily defined realm than our own hard-edged terrestrial one. They inhabited what ancient Oceanic peoples called "the Dreamtime," a gauzy, blissful place located somewhere between our generally-agreed-upon reality and any number of sublime alternate states.

Certainly, dolphins have a laundry list of capabilities that qualify as magical. They can see with their hearing, deploying biological sonar to effectively produce X-ray vision: dolphins can literally see through objects. They know when another dolphin—or a human being—is pregnant or sick or injured. Their echolocation skills far outmatch the most sophisticated nuclear submarines; scientists suspect they can even use them to determine another creature's emotional state. They can communicate at frequencies nearly an order of magnitude higher than anything humans can discern, and navigate electrical and magnetic fields imperceptible to us. They can stay awake and alert for fifteen days straight.

Recently, scientists have marveled at dolphins' healing abilities, which include infection-resistant, pain-free, hemorrhage-proof rebounds from even the deepest wounds. In a letter published in the delightfully named *Journal of Investigative Dermatology*, researcher Michael Zasloff, MD, described the process as mysterious and miraculous, likening it more to regeneration than repair. "Despite having sustained massive tissue injury, within a month the animal will regain its normal body contour," he explained in an interview. "A chunk of tissue maybe the size of a football will have been restored with essentially no deformity." He also surmised that dolphin tissues might contain "the long-sought natural morphine that we've been looking for."

The dolphins' evolutionary path is itself a preposterous feat: their predecessors were land mammals that resembled small, hooved wolves. After an interlude in swamps and coastal lowlands, these fledgling aquanauts moved permanently into the water. Over the course of twenty million years (give or take a million or two), their limbs turned to fins, their shape became streamlined for swimming, their fur turned to blubber, their nostrils migrated to the top of their heads—in other words, they developed all the equipment needed to master undersea life. They aced it, too: dolphins have perfectly hydrodynamic bodies. They swim faster than physics would seem to allow, given the density of water and the amount of muscle they have. Their bodies are so ideally adapted for speed, navigation, plunging into the depths, and keeping warm that it's hard to imagine improvements.

But while it's tempting to project onto dolphins all the superpowers we wish we had ourselves, I knew (on an intellectual level, anyway), that these were creatures who have it in them to be cranky and withdrawn and have their own version of a bad day. It is now widely known that dolphins don't always act like the gentle, perma-smiling unicorns they're often made out to be; their range of less-than-cuddly behaviors is actually quite complete. In fact, despite the vast differences between our two species, possibly the most startling thing about dolphins is how inexplicably they resemble *us*. "It's like dolphins and whales are living in these massive, multicultural, undersea societies," said Hal Whitehead, a marine biologist from Dalhousie University. "Really the closest analogy we have for it would be ourselves."

In any group of dolphins you'll find cliques and posses, duos and trios and quartets, mothers and babies and spinster aunts, frisky bands of horny teenage males, wily hunters, burly bouncers, sage elders—and their associations are anything but random. Dolphins are strategists. They're also highly social chatterboxes who recognize themselves in the mirror, count, cheer, giggle, feel despondent, stroke each other, adorn themselves, use tools, make jokes, play politics, enjoy music, bring presents on a date, introduce themselves, rescue one another from dangerous situations, deduce, infer, manipulate, improvise, form alliances, throw tantrums, gossip, scheme, empathize, seduce, grieve, comfort, anticipate, fear, and love—just like us.

The Hawaiian dolphins were like some ancient tribe I had stumbled upon, and though I didn't understand their lineage or their language, they had somehow communicated with me. More important, for reasons I could not say, the spinners made me feel better. They took the edge off my sadness. During the moments I revisited them in my mind, the dolphins made me feel happy again.

≈≈≈

Once I started paying attention to dolphins, I began to notice them everywhere. They were no strangers to the headlines, and extremely popular on the Internet. I read stories about dolphins helping salvagers locate undersea buried treasure, dolphins saving surfers from imminent shark attacks, dolphins recruited as soldiers by the U.S. Navy. While scientists argued about the existence of animal culture, dolphins and their close relatives, whales, were observed making babysitting arrangements among themselves, congregating for a funeral, and calling one another by name. *The Guardian* reported that a beluga whale named Noc had, after seven years in captivity, begun to mimic human speech. Belugas, members of the toothed whale family along with dolphins, have been nicknamed "the canaries of the sea," for their expressive vocals. Among other things, Noc had apparently demanded that a diver in his tank get out of the water. In their science paper describing this event, the authors referred to "other utterances" that sounded like a "garbled human voice, or Russian, or similar to Chinese." The whale had been insistent, the paper revealed: "Our observations led us to conclude the 'out' which was repeated several times came from Noc."

We've long known that dolphin brains are impressive, bigger even than the brains we consider the gold standard: our own. Yet science still searches for answers to what the dolphins are *doing* with such metabolically expensive machinery (and, for that matter, what human brains are really up to). No creature would cart around a big brain if this heavy artillery wasn't in some way essential for its survival. A clue emerged when dolphin brains, like humans', were found to contain von Economo neurons: specialized cells that relate to higher notions like empathy, intuition, communication, and self-awareness. Interestingly, dolphins have far more of these neurons than we do, and they are thought to have developed them 30 million years ago, about 29.8 million years before *Homo sapiens* swung their first clubs.

Even so, despite the similar heft of our gray matter and our shared ability to express irritation, I was surprised to learn that the dolphin genome, sequenced in 2011, bears a striking resemblance to our own. When researchers compared the dolphins' gene mutations to those of other animals, they found 228 instances where the dolphins had done something smarter, evolving in ways that revved up their brains and nervous systems. These adaptations aligned them more with humans than with any of the other species tested, even those which were more closely related to dolphins. Having been around for so much longer than we have, dolphins had also developed some nifty tricks: one of their responses to type 2 diabetes, for instance, is to internally flip a biochemical off-switch and block the disease's progression.

While scientists made news with their dolphin findings, the animals also caught the attention of the film world. *The Cove,* a movie about a barbaric dolphin hunt in Taiji, Japan, riveted audiences, and went on to win the Oscar for Best Documentary in 2010. Each year, the movie showed, local fishermen conduct this hunt, driving pods of bottlenose, striped, white-sided, Risso's dolphins—any dolphin they can catch, basically—into a narrow cove, then netting off the entrance and killing the animals with gaffs and long-handled knives. Getting rid of as many dolphins as possible—whom they view as competition for what few fish remain in the vacuumed-out oceans—the fishermen claim, is a vital matter of "pest control."

Most of the captured dolphins end up in Japanese supermarkets and restaurants (though the meat is highly contaminated with mercury and other toxins), but some do not. Younger females and calves are separated out, examined by trainers and dolphin brokers, and then sold to marine parks for six-figure prices. Every year, the hunters kill or sell thousands of dolphins.

Once the Taiji hunt got dragged into daylight, celebrities like Jennifer Aniston and Woody Harrelson and Robin Williams spoke out against it, drawing even more attention to the secretive little town. Sadly, Japan wasn't the only place dolphins were dying en masse—they were washing up on shores all over the world. Scientists have scrambled to find a possible cause for the global die-off, but pinning down a single problem is hard—there are so many. Over in California, the bottlenose dolphins were suffering from gaping skin lesions. Across Europe, striped dolphins have washed up emaciated and riddled with herpes, their immune systems hopelessly compromised. In Florida, dolphins fall victim to runaway cancers. Dolphins everywhere, from Australia, to North and South America, to Tahiti, are so laden with industrial pollutants—pesticides, heavy metals, flame retardants, carcinogens of the most noxious kinds—that their bodies are disposed of as hazardous waste. If this chemical onslaught weren't enough, the acoustically sensitive dolphins also contend with a clanging underwater mayhem of drilling, ship engines, oil-rig construction, explosives, and submarine sonar that can blast sound across entire ocean basins; this bombardment harasses millions of animals, and can even kill them. "The future for dolphins is a lot gloomier than their smiling faces suggest," the magazine *New Scientist* wrote in an editorial.

If reading about these travails makes you upset, you are not alone. It's not just the idea of dolphins in trouble—dolphins in general strike a deep emotional chord in most people. On some level, however vaguely at times, we seem to know how connected we are, the dolphins and us, and how inevitable it is that we share the same fate. Rigorous science balks at the notion that these animals affect us so profoundly because of some innate spiritual connection, but that doesn't make us feel it any less. Anyone who's ever spent time around a dolphin, any dolphin, faces abstract, philosophical questions such as these posed by marine biologist Rachel Smokler: "Do [dolphins] have the same powers of reasoning that we have? . . . Do they feel love and hate, compassion, trust, distrust? Do they wonder about death? Do they have ideas about right and wrong and accompanying feelings of guilt and righteousness? What could they teach us about the oceans? How do they feel about one another? What do they think about us?"

Regardless of the dolphins' allure, few people feel as close to them as Sharon Tendler, a concert promoter from London who became the first human to officially marry one. Tendler and her groom, a thirty-five-year-old male bottlenose named Cindy, had courted for fifteen years at the Eilat Reef resort town in southern Israel. The bride wore a flowing white gown, a veil, and a headdress of orchids as she kneeled dockside to kiss Cindy, who accepted her gift of mackerel. Though dolphins do have a reputation as ladies' men, showing an unusual amount of interest in interspecies amour, Tendler declared that the marriage would remain unconsummated. "I'm the happiest girl on earth," she told the press, adding: "I am not a pervert."

≋

Clearly, dolphins are charismatic enough as plain old wild animals; they don't need to be angels or gods or spiritual guides in the bargain. Undeniably, though, they get nominated for these positions. Walk through any New Age bookstore and note the dolphins per square foot; you'll find them on bookmarks and posters and stickers, glittering in 3-D notecards and tinkling on wind chimes, adorning CD covers and T-shirts, leaping across the covers of countless journals. (Note, too, that in the forest or the savannah or the jungle, even the most impressive beast is not usually mistaken for husband material.)

What *is* it about dolphins? Why do we obsess about them so? As far back as anyone can cast in history, there is evidence of a unique bond between us. The Maoris and Aboriginal Australians and Pacific Islanders, the Greeks and Romans: Odysseus, Poseidon, Apollo, Aristotle, Socrates, Plutarch, the Plinys Younger and Elder, the Emperor Augustus—they were all dolphin crazy. Actually, everyone was. Dolphins were painted on palace walls, sculpted into statues, stamped on gold coins, tattooed onto bodies. In ancient Greece, apparently, dolphins had the same rights as people. Perhaps even greater rights: while it was considered perfectly all right to snuff your disobedient slave, to kill a dolphin was equal to murder. Our relationship might even have included literal conversation. In his 350 BC *Historia Animalium,* Aristotle wrote, "The voice of the dolphin in air is like that of the human in that they can pronounce vowels and combinations of vowels, but have difficulties with the consonants."

The image of a sea creature poking its head above the water to speak to us is like something out of *Alice in Wonderland* or the latest Pixar masterpiece—an irresistible thrill. Theoretically, anyway, dolphins have the brainpower and the communication skills to do this, and so they occupy a singular place in our imaginations. They take us back to our earliest years, to that little blip of time when we believed that we *could* communicate with other creatures, because there was no separation between their world and ours. "When we were children we wanted to talk to animals and struggled to understand why this was impossible," the naturalist Loren Eiseley wrote. "Slowly we gave up the attempt as we grew into the solitary world of human adulthood." This loss of hope, Eiseley pointed out, is a very sad thing.

Dolphin intelligence may come in a different package than human intelligence, but a thread of awareness connects us. It's an ephemeral link, an ember almost. Although we can't easily define it, we seem to long for it. In some deep-seated way, we hope to find other wisdom, other guidance—*others.* It's the reason we point telescopes toward the stars, and wonder if there's anyone out there who wants to talk to us.

Even the slightest possibility that the answer might be yes both terrifies and enthralls us. Given our curiosity about the bigger questions, our hunger to know more about the purpose and scope of our lives, it's really not that unreasonable to wonder if behind their Mona Lisa grins, dolphins might be in on some good cosmic secrets.

≋

When I think back on it now, my swim with the spinners in Honolua Bay was an experience as mystifying as it was uplifting. Who *were* those creatures? It's been said that humans are the only animals who believe the stories they tell about themselves—but what about the dolphins? What is their story? And what about those haunting sounds they made? Their whistles and clicks and squeals seemed to me like a liquid symphony, a communiqué from another realm, a galaxy of meaning conveyed in a language that defied translation. When I saw the pod, I felt joy. I felt awe. And I felt the slightest bit frightened, though the dolphins were not scary. I felt their beguiling mix of mystery and reality; I felt a sense of bottomless wonder.

The one thing I didn't feel was alone.

CHAPTER 1

THE MEANING OF WATER

Hawaii's Big Island is huge and low-slung, the product of five feisty volcanoes. The youngest of the visible islands in a 3,600-mile-long undersea mountain range, it sports dramatic whorls and swirls of lava, punctuated by tufts of pale green fountain grass. Steering my rental car out of the Kona airport, it was impossible to miss massive, crouching Mauna Loa, the world's largest active volcano, or fail to note the lunar starkness of its surroundings. Even a sparkling new Target store built on the lava fields below couldn't disguise the elemental nature of the place, the clash of molten rock and sea. Anyone visiting this island must confront the daunting truth that its story sprawls more than 500,000 years back, far beyond prehistory—and yet it has barely begun.

I drove south toward the town of Kailua, an infinity of ocean to my right, fronted by more charred black lava. To my left the clouds hunkered low, exposing Mauna Kea's summit. If you traced this volcano down to the bottom of the ocean, you'd be measuring a mountain higher than Everest; most of Mauna Kea happens to be underwater. At its peak I could make out the glinting domes of its thirteen telescopes, world-famous perches from which astronomers study the night sky. Here in the middle of the Pacific Ocean, far from the diluting lights of cities, they can more clearly glimpse the heavens than anywhere else on the planet—the billions of stars, the planets and exoplanets and their many moons, the solar systems with their nebulas and starbursts and voids, the asteroids and comets and supernovas, our sun and other suns, all the question marks floating out in space. "Hawaii is Earth's connecting point to the rest of the Universe," the Mauna Kea Observatories' Web site proclaims.

But I had come to the Big Island to look down, rather than up. It was the depths that interested me, the universe below the ocean's surface; and it was there, underwater, that I hoped to find my own connecting point: these waters teemed with dolphins. There were spinner multitudes here, megapods hundreds of animals strong, and each morning they appeared so reliably in certain bays along this coastline that an entire community of people had formed around them. They called it "Dolphinville," though it was less an actual place than a shared state of mind. "There are approximately 200 of us, and we live in separate homes along a 30-mile stretch of the Kona Coast, connected in spirit to each other," I read, in a description of the group. "Many of us who live here have been called by the dolphins. We have become like a family who swim, meditate, and work together. Swimming among the dolphins day after day, we are in deep communication with them." Accompanying this description was a group photo with dozens of suntanned, smiling, athletic-looking people who seemed like they were having a hell of a lot more fun than anyone I knew back on the mainland. Nobody looked demonstrably crazy.

I was enchanted to learn about Dolphinville, figuring that even if it was a wacky cult it would be the first one I'd heard of that involved three- and four-hour open water swims. My interest was further piqued when I found out that Dolphinville's creator (or pod leader, depending on how you look at it) was a woman named Joan Ocean. A New Jersey–born psychologist turned New Age dolphin guru, Ocean had, by her own estimate, logged over twenty thousand hours with wild dolphins. When I'd e-mailed to see if I could come to the Big Island and join her for a swim, she'd not only responded yes, but also invited me to stay at her house.

I was anxious to get back in the water. After my first encounter with the spinners I'd reshuffled my priorities, making time and space to explore the strange, enduring, occasionally tragic, and often wonderful relationship between humans and dolphins. My reasons? Because the idea of such a quest brightened my life—and my father would have encouraged it. Because I was too curious *not* to follow wherever the dolphins might lead me. Because I wondered if there was some greater understanding possible at the place where our world and theirs intersect, and what that might mean for both of us. And because I really wanted an answer to this one confounding question: Why had a mere ten minutes in the dolphins' presence been such a soul-shaking experience?

Since the sixties, dolphin research had proceeded at a heady clip, revealing much about the animals' biology and physiology and cognition, but the more we learned, it seemed, the more we needed to make sense of our findings, not only with our minds but also with our hearts. I knew the questions I was asking couldn't be answered solely by consulting scientific papers or marine mammal textbooks, fascinating as those were. My answers could only be found out there, in the ocean.

≈

"I call what I do 'participatory research,'" Joan Ocean said, standing barefoot on the harbor dock, flashing a high-voltage smile. We were waiting, along with about twenty other people, for a dive boat called *Sunlight on Water* to start up its engines and take us out in search of local dolphin pods. The morning was bright and wide open, the sea an inviting marine blue. While everyone else carried the standard-issue dorky array of snorkeling gear, Ocean held a pair of free-diving fins with sleek, attenuated blades. Among the clot of tourists, she stood out. Her long hair was a tangle of silver and blond, but it was her eyes you noticed first. They were a laser shade of aquamarine, animated with an excited, kid-like energy. She wore a diaphanous floral cover-up over a black bathing suit, and sparkly pink toenail polish. Ocean, seventy-four, was a great-grandmother, but she looked oddly ageless. "If you wanted to learn about any culture, you would move in with them if they allowed it, and observe them and then try to be like them," she continued. "So that's what happened with the dolphins. The first twelve years I lived here, every single morning I was in the water. You just learn a lot."

When Ocean first took up with the animals back in the late seventies, she hadn't even known how to swim; she was forty-five when she took her first stroke. But she quickly made up for lost time, regularly swimming miles at a stretch. The new saltwater world she had entered, she said, "was like being on another planet, only better." In the years before she became immersed in dolphins, Ocean had spent gritty time in the human realm, counseling abused children, delinquent teenagers, battered women, fractured families. She had empathy, she cared, and she tried with everything she had to coax her clients out of destructive patterns, only to watch in frustration when they repeated themselves. She prayed. She meditated, asking for better ways to help people. What she got back in her mind's eye were images of dolphins (and the occasional whale). She began to seek them out: bottlenoses in Florida, orcas in British Columbia, spinners in Hawaii, botos in Brazil. Every encounter with them lifted her spirits—why, she wondered, wouldn't this be the same for others? She felt the animals were communicating messages of love and wisdom, imparting information that we desperately needed to hear, albeit telepathically. From that point on, Ocean's purpose was clear: introducing people to cetaceans.

The resident Kona pods, she explained, lived in what was known as a "fission-fusion" society. Essentially, the mass of dolphins moved in fluid, constantly changing groups, much like people milling around at a cocktail party. This was a sophisticated arrangement, uncommon in nature, requiring the animals to recognize one another, form bonds, trade favors, recall past associations, and get along in unfamiliar circumstances. Scientists had wondered what factored into a dolphin's decision to leave one group or join another; they discovered that dolphins bopped between social clusters for the same reasons humans might. A teenage dolphin swimming along with his mother, for instance, might defect to a band of teenage dolphins who were having raucous fun; females with calves liked to hang out together; mating pairs were mostly interested in one another. In dangerous situations or tricky hunting conditions these subgroups would merge back together into one larger pod. When the heat was off, they would drift apart again. To researchers it seemed likely that the spinners along this coast all knew one another, at least on an acquaintance basis.

Ocean, as she was quick to point out, was not a scientist. She did not publish academic papers or bear any special set of credentials. "I didn't choose that particular path," she said. But she knew these dolphins. She had observed their society in action, day in, day out, for twenty-six years. She knew their habits and their quirks, their likes and dislikes and the body language that expressed them, their fondness for playing with the leaves that fluttered down from hala trees into the ocean. She recognized individuals, and charted the company each dolphin kept; she noted when one animal showed up with a cookie-cutter shark bite, a missing divot of flesh, or another bore signs of a boat propeller strike. She knew their rhythms. Each morning after dawn, pretty much like clockwork, the spinners would return from a night of feeding in deeper waters and move into the island's shallower bays, circling slowly, schmoozing, and in general taking it easy until mid-afternoon, when they'd begin to commute offshore again. This nocturnal hunting schedule was for good reason: at night, schools of fish, squid, and shrimp rose from the depths and became a dolphin smorgasbord.

It would be hard to find someone more jazzed about dolphins than Ocean, I thought, or someone who had created a more dolphin-centric life. While other people hacked their way toward the regional sales record or Ivy League admission or any number of goals along various well-worn paths, Ocean had been doing things like swimming with pygmy sperm whales in the Mexican moonlight, frolicking with pink river dolphins in the Amazon, and examining dolphin hieroglyphics inside the Great Pyramid. Hers was the opposite of a normal career trajectory, and because of that it had been fraught with questions like "Can I pitch my tent here?" and "Who will buy the groceries?" But a funny thing happens when you refuse to compromise your heart's desires: after years of contending with all manner of uncertainties and obstacles, life yields and pours itself into a you-shaped mold, one that has never existed before. After decades of out-of-the-box "participatory research" with dolphins, and quite against the odds, Ocean had forged a prosperous, even settled, existence. Dolphin lovers worldwide sought her out, flying to Hawaii to attend her weeklong workshops, which were often sold out months in advance.

Sunlight on Water floated serenely in its slip while its passengers fumbled with flippers, cameras, sunblock. A sporty Hawaiian man in wraparound sunglasses came charging down the dock and wrapped Ocean in a bear hug. It was the captain and owner, Mike Yee, affectionately known as China Mike. "Me and Joan we go way way way way back!" he announced. "Ha ha! I've been taking this young lady out for a quarter century!" Yee gestured exuberantly to the people milling around. "Hey, alright, since everyone's here I'm gonna do a Hawaiian ceremony to begin the day," he said, the words tumbling out in a monologue. "Anytime people get together here in Hawaii we always begin with some ceremony. It's part of our tradition. Doesn't matter if you're making a boat trip or it's a birthday party or, you know, any of that kind of stuff, ya? This is a Hawaiian wind instrument I've made. It's called a *pu*. Ya? It's made of *ohe* bamboo. What? This is not school, honey. You don't have to raise your hand. Anyway, I'm not answering questions now. So I'll blow this four times to begin the ceremony. When I start I'm always facing east, acknowledging *kala*, the sun, the life force of our planet. Ya, you know, we figured it out two thousand years ago: if the sun doesn't come up, not too much happens here on good old planet Earth."

Yee blew into the *pu'ohe*, which sounded low and mournful, like a quavery foghorn. Everyone bowed their heads. "We ask for blessings on our boat today," he said. "We send out our love to our *aumakua*, our spirit guides, all the dolphins and whales on our planet. And we ask permission that we may swim with you today, and experience all the joy and happiness and knowledge and wisdom that you bring to our world. So it is, *amama*." He laughed. "Okay, you're covered."

Today was Yee's day off, he explained as everyone boarded, so another captain named Jason would be taking the wheel, assisted by first mate, Dusty. Jason and Dusty were young, sun-scoured guys who looked like they would rather be surfing than spending the day babysitting the besnorkeled masses. As we chugged out of the harbor, Ocean listened politely as Dusty held forth about dolphin swimming etiquette in a laconic, stoner patois: "So yeah, try to be mellow. The more mellow you are, the more calm and wise you're putting out there, the more they're gonna be into you. That overhead Michael Phelps kind of swimming? You don't want to do that around the dolphins. They think that is aggressive behavior. It will make them not want to hang out with you. They're very awesome."

"How long do they live?" a stout woman asked in a Texas drawl.

"Uh, twenty, maybe twenty-five years—"

"Well, that's in captivity," Ocean cut in. "We don't really know how long they live, but we think it's probably around seventy, seventy-five years. I have all the notes from the first researcher here, Ken Norris, years and years ago. The dolphins he was swimming with are still here, a lot of them. So we have some pretty good evidence that they live longer than people think. And whales can live to be a hundred."

Her voice was drowned out as the engines powered up and we motored south, toward a series of bays the dolphins frequented for their daytime repose. "Would anyone like a muffin or perhaps some fruit?" Dusty yelled, above the din. "Coffee? Snacks?"

Ocean and I sat together by the stern, ready to launch ourselves into the water whenever the dolphins appeared. The Kona coast slid by in the clear morning light, spray from the wake casting tiny diamonds into the sky. It was an ideal day for a reunion with the spinners—and I knew they would be here. We were headed into the center of the spinner universe, undeniably the best place to go if you wanted to be sure of finding them. The high dolphins-per-square-mile ratio of this island had not gone unnoticed by scientists. Along with the big population of spinners, they could reliably find even more exotic dolphin species here, deepwater creatures like pilot whales, false killer whales, rough-toothed dolphins, striped dolphins, dwarf killer whales, Risso's dolphins, and melon-headed whales. Bottlenose and spotted dolphins regularly showed up, too.

The man whose name Ocean had cited, Ken Norris, was a giant of dolphin science. He had pioneered studies of Hawaii's spinner population in the late 1960s through the mid-1980s. During that time, Norris, who revealed many basic facts about dolphins, called his study subjects "the most mysterious of fauna on the planet." He ended up proving that dolphins were masters of the world of sound, that they used their sonar to paint exquisitely nuanced pictures of their surroundings, discerning details down to the molecular composition of an object. Even blindfolded, a dolphin could tell the difference, for instance, between a sheet of copper and a sheet of aluminum. These extraordinary abilities fascinated Norris, who posed a challenging question: "In an ocean full of dullards, what good is such a brain? Certainly complicated nervous machinery is not needed for concourse with jellyfish, sea cucumbers and sponges." He'd had a glimpse of what the dolphins could do—the next step was to figure out why.

Norris's research began just seventeen miles south of here, in Kealake'akua Bay, a flawless inlet ringed by sea cliffs. (Translated from Hawaiian, its name means "Pathway of the Gods.") Along with its abundant dolphins, Kealake'akua is known for its memorable snorkeling, storybook sunsets, sacred status, and infamous history: it was the place where Captain James Cook landed in 1779 and then, in a fracas with natives over the theft of a boat, was stabbed to death.

"The Kealake'akua Bay topography was perfect for dolphin work," Norris wrote in an essay titled "Looking at Wild Dolphin Schools." "A magnificent, nearly vertical 500-foot cliff loomed over the almost-always calm and clear semicircular bay. A group of dolphins rested in the bay nearly every day. One could look down almost on top of those resting dolphins, and their behavior could often be seen in toto as they moved offshore, until they faded from view into the gray disk of the sea."

Norris and his colleagues set up camp on a bluff, peering down at the dolphins through telescopes. When this vantage point proved inadequate, Norris jury-rigged an underwater observation vehicle, known as the SSSM: Semisubmersible Seasick Machine. Later, the SSSM was retired in favor of other contraptions with improved viewing capabilities, all of them dreamed up to better observe the spinners. "We need to see other dolphins as a dolphin sees them," Norris wrote.

Ocean, too, had set out to do exactly that, and while the scientists went about their work, amassing and quantifying and analyzing data, she continued to swim in her backyard—Kealake'akua Bay. From the small house she rented on its rocky shores, she would launch at dawn and venture out to meet the spinners. She swam with them on cloudy choppy days and stormy rainy days and flat overcast days and days when the sun blasted down and turned them into flippered silhouettes. At times other people joined her, residents of Dolphinville or the occasional intrepid tourist. But much of the time, even far offshore, Ocean swam alone.

"I've always been guided by the dolphins," she told me. "I could totally trust them—they would never crash me into coral or take me out to sea. I could swim eye-to-eye with them for hours. I never had to look up, I could breathe through my snorkel." Swimming like this for extended periods, Ocean said, vaulted her into "an altered sense of awareness," one in which her mind slowed and her worries slipped away, where her senses were heightened and she felt attuned to the slightest movement, even something as subtle as a fish nibbling on coral. "As time went on and I became completely at home in the water," she said, "I began to understand their language."

As Ocean spent endless hours among them, the dolphins rewarded her with some rare and beautiful sights: one morning, for instance, she watched five spinners give birth simultaneously, the babies corkscrewing out of their mothers tail-first, then wobbling to the surface to take their first breaths. Witnessing this made her wonder if the dolphins could decide, within reason, when they wanted to deliver—because statistically, five babies appearing at once was almost certainly not random. That possibility, in turn, led her to muse: Could dolphins also choose the moment of their own deaths? This suggestion had been raised before, though no one had even come close to proving it. (Scientists did know that dolphin erections happened at will; they could pop it out like a kickstand or retract it neatly whenever they wanted to. What other body functions did dolphins have under their own control? It was an intriguing question.)

The boat slowed as Jason drove into Kailua Bay, a long sweep of shoreline dotted with hotels, condominiums, stores, and restaurants, connected by an oceanfront promenade. The water was calm here, making it a favorite spot for swimmers, paddleboarders, and kayakers. Every October, this bay boiled white with spray when two thousand Hawaii Ironman competitors stampeded in for their 2.4-mile opening swimming leg, and for this Kailua was famous. I looked out at the Jet Ski rental concession that floated atop a squat platform festooned with palm trees; at the moment, mercifully, no one was there. At least five other dolphin-watching boats idled; snorkelers could be seen everywhere, kicking across the surface, draped across cylindrical, fluorescent-colored floats called noodles, and in general bobbing around. It was a bustling bay, hardly the place you'd go for a pristine wildlife experience. But the spinners, apparently, had chosen it. Fins broke the surface all over. Once my eyes adjusted to the sunlight's glare, I realized the extent of the gathering: there were hundreds of dolphins in here.

Jason cut the engines. "Pool's open!" he yelled, and with that, Ocean and I slid into the water. Immediately she swam off, guided by internal compass in the direction of the pods. I followed. The water was warm, an enveloping blue cocoon, and I could see all the way to the seafloor, sixty feet down and speckled with reefs. Midsize fish cruised near the bottom, cast in monochrome tones by the depths. Only the dimmest tinge of yellow was visible as the spectrum ebbed. I was startled by the contrast between the peacefulness right below the surface and the scrum of activity above, as people clambered off and onto boats. The dolphins, however, were nowhere in sight.

I had read about the shy nature of the spinners, who prefer to be the ones who initiate a meeting. When you dump a hundred snorkelers into their midst they tend to back off, regroup, and then decide how much contact they want. Unlike bottlenose dolphins, who often initiate play with humans, the spinners are reserved, almost standoffish. "I always tell people, don't swim after them," Ocean had warned. "You'll just wear yourself out. They'll circle outside of all the boats and all the people. They don't go away—they're circling. But in the beginning when everyone gets into the water, it's like mayhem. The dolphins will sonar the people and wait for them to calm down."

Adjusting my mask, I scanned the other snorkelers I could see in the bay. About twenty feet away, a guy in neon green surf trunks swam frantically holding a GoPro camera mounted on a long pole, legs pumping, churning away in his fins. His head snapped back and forth, surveying the ocean impatiently. The GoPro swept through the water like a 9-iron. Facebook awaited, Instagram, Twitter. *Where are the goddamned dolphins?* his body language said. He aggravated me; I suspected the dolphins felt the same.

Then, far below, I made out their shapes. A dozen spinners glided by, fin to fin. They were thirty feet down and all I could do was admire their hazy outlines from above, but still it was enough to captivate me. These dolphins were part of a larger group, and when the rest of the pod flowed past, every person in the vicinity shot off after them. Even the most sluggish dolphin was moving exponentially faster than flailing snorkelers could match, however, so this was a futile strategy. "They're better swimmers than we are," Dusty had observed dryly, adding that, "If you chase them, all you'll see today is a bit of receding tail."

Ocean took off in the opposite direction, and I followed her as she dove to twenty feet, kicking strongly. Almost instantly, three dolphins appeared in front of her. Ocean dove deeper, tucking in behind them as if part of their posse. The dolphins swept low and then arced toward the surface, breaking through to breath before submerging again. I was struck by the synchronization of their movements, as precisely timed as an Olympic event, and yet they could not possibly have been anticipated. Somehow, each dolphin just knew what the other dolphins were about to do, and they moved effortlessly as one.

We spent the morning in Kailua Bay because the dolphins stayed there too, orbiting at various perimeters. Every so often they'd cut diagonally through the bay, allowing us a few snatched moments in their company. There was a similar rhythm to the encounters; Ocean referred to them as "drive-bys." Still, it was entrancing to watch such a sprawling pod of dolphins, even if they weren't particularly interested in us. If I'd been hooked up to a heart-rate monitor, it would have shown a distinct slowing: the spinners had the same Zen effect on me as before. At times I found myself drawn along with them, deeper down than I usually dived and lost in a blissed-out daze, surrounded on all sides by narcotic blue. It took a conscious effort to leave them, to bow to my need for oxygen and head for the surface, and every time I did so I felt an ache of regret.

At one point a manta ray sailed by, its great black kite of a body undulating over the seafloor, its whip of a tail trailing elegantly behind. I stared at it for so long that I almost missed the two dolphins who were hovering behind me, looking at me with the same intent interest I had focused on the ray. These were not spinners, I could tell immediately. They were bigger, thicker, and far less demure. One of them moved closer and swung his head toward me, and I felt him zap me with a burst of sonar, then heard him let loose with a barrage of eerie creaking noises, like a door swinging back and forth on rusty hinges. Our eyes met, and he nodded his head repeatedly, in an almost agitated fashion. It felt as though we were having a heated conversation, but I had no idea what it was about.

Having sized me up, the alpha dolphins swam past me to a man on a paddleboard, nearby. He was standing, barely upright on shaky legs, and he held his paddle backward as he tried hard not to pitch into the drink. The dolphins circled him, verging on aggressive, following as he lurched back toward the harbor. When the man noticed them he became even more rattled and dropped onto his knees, losing the paddle and gripping the edge of his board with both hands. I didn't usually think of dolphins as intimidating creatures, but these two reminded me of that possibility. Earlier, Ocean recounted how she'd once been whacked so hard on the leg by a dolphin's tail that she was unable to walk afterward; on another occasion a bottlenose had clamped its jaws on her calf, staring at her defiantly and refusing to let go. "People think, 'Oh, dolphins, all sugary sweet and loving,'" Ocean said, with a laugh. "And they are—but they also have such power."

As the morning unfurled, the bay gradually emptied of snorkelers and the spinners seemed to relax, swimming slower and coming closer. This was their version of sleep, after all—and who can doze in the middle of Grand Central Station? Dolphins never actually shut both eyes and nod off, like we do. Unlike us, they are conscious breathers. Whenever they take in oxygen it's a decision, not an autonomous body function, which makes sense if you think about it. They are air-breathing mammals who happen to live in the sea. If a dolphin were knocked out underwater and his body continued to try to suck in air, he would drown. Dolphins themselves seem to understand this: when a dolphin loses consciousness his podmates will lift him to the surface, holding him up there until he's revived.

Because each breath is intentional, the animals have to keep swimming, stay vigilant, maintain system operations at all times. Physically, this is a tall order. Imagine if instead of snoozing under a fluffy duvet, you were required to run a slow marathon or cycle an easy hundred miles in your sleep. This would destroy us, but dolphins are able to do it because they operate the two hemispheres of their brain independently; while one side runs the show, the other side rests. It's a formidable juggling act, the kind of evolutionary genius that happens when a species sticks around for tens of millions of years. Even when they nap, dolphins are at least half awake. So it's no wonder the spinners weren't always rambunctiously playful when they pulled back into these bays.

By swimming together, they look out for one another, too. Their strategies, their physiologies, their routines—everything about them works for the continued health of the group. For dolphins, there is strength in numbers and safety in numbers and, from what I'd observed, joy in numbers. Ken Norris had arrived at that same conclusion: "I view the school as the matrix within which their lives are played," he wrote, in a 1978 letter to another scientist, "and hence the structure that contains everything they are." The dolphins' entire existence, Norris came to believe, was lived in the collective.

After a while I became cold and returned to the boat. Most of the other passengers were already on deck, eating pineapple slices and comparing dolphin stories. "I feel like they raised my frequency, you know?" a red-haired woman with ambitious tattoos told an older woman wearing a sunhat tied under her chin. Next to them, a boy who looked to be about ten years old sat sullenly, ripping into a jumbo bag of chips. He wore cubic zirconia studs in both ears and a bracelet that said I LOVE BOOTY.

Ocean climbed back on board, and peeled off her fins. "Did you see those two spotted dolphins?" she asked. Even when she was excited, Ocean's voice was soft and low-pitched, just slightly raspy. When she said the word "dolphin," she drew the syllables out for an extra beat—"*dahhhlllfin*"—as though she especially loved the sound of them in her mouth. I told her I had definitely seen the bigger dolphins, and that they had given me the once-over at close range. The effect, I added, was somewhat unnerving. "Oh, yeah," Ocean agreed. "The spotteds are cheekier. They'll even treat people in the water as a challenge. They'll swim right up to you and clap their jaws. And the spinners would never do that. They're always kind. We call the spotteds 'the Bikers.'" Dusty, standing beside us, nodded. "The spinners are like, 'Heyyyy, what's going on?'" he said, "and the spotteds are like, 'HEY WHAT'S GOING ON!!!'"

Suddenly, off the bow, a spinner shot out of the water, torpedoing into the air and whirling at least 900 degrees. Another dolphin followed seconds later, executing a full aerial somersault. Everyone cheered and clapped. "That right there was a blessing!" the woman from Texas yelled. Ocean smiled, towel-drying her hair. "Why do they spin?" I asked. "Is it about dominance or mating or—"

"Fun," Ocean said. "It's more like fun. They seem to do a lot of things for fun." But it also had to do with signaling to one another and removing remoras, she added, the suckerfish that attach themselves to dolphins' skin and then hitchhike along feeding on castoff bits of fish or plankton or anything else that came their way. "I had a remora on my leg once," she told me. "It was the cutest little thing. But it hurt!" I wondered how long you had to swim around before remoras showed up and attempted to paste themselves to your body; the short answer was a very long time. "Do you ever feel as though you could just stay down there?" I asked. "Find a pod and just keep going?"

Ocean laughed, and nodded: "All the time."

≈

The road to Ocean's home, Sky Island Ranch, starts out as a busy coastline highway then narrows to a local thoroughfare before it veers up a hillside, dwindling to a single lane cut through jungly cloud forest until it widens to reveal a gray wooden house with a grooved red-metal roof. It's a place built to endure rain, sheets and torrents and lashings of rain, surrounded by a riot of vegetation. Everywhere you look there are flowers and ferns and trees and vines and bushes bursting out of the ground, as though intent on seizing control the moment anyone steps away. "We never expected to buy property," Ocean said, as she drove past the gate and up the driveway. "We wanted to live like dolphins and not own anything. But then a friend who was a real estate agent said, 'You've got to see this house. I think it's perfect for you.' And—boom!"

Ocean lives here with three miniature horses, two donkeys, a trio of cats, and her longtime friend and collaborator, Jean-Luc Bozzoli. Bozzoli is an artist who paints intricate, ethereal dreamscapes, hallucinogenic scenes from far-off planets and other dimensions, many of which feature dolphins. He and Ocean met in the seventies and bonded over their shared New Age convictions, beliefs that ranged from the reasonably mainstream (dolphins are highly evolved, intelligent creatures) to the far fringes (dolphins' sonar vibrations activate dormant human DNA so we can receive encoded messages from other planets).

On this spectrum I fell somewhere in the middle, filled with questions about the nature of reality and inner space, intrigued by the likelihood that human existence was more expansive than our minds typically grasped, curious enough to have flown down to Brazil once to visit a spiritual healer. But if you asked me what I *really* believed, which convictions I held firmest and strongest and would stand behind no matter what, I could only offer something terrifically vague: that I believed life was more magical than we usually allowed for. That I believed there was a God, though I had no idea what that God looked like, other than he or she or it must exist in the grand architecture of the universe down to the finest details of nature, a nature that humans are not separate from but beautifully and inexorably part of. That miraculous, spellbinding, challenging things happen continually, beyond what our minds and senses can process, governed by an intelligence that we can't see and don't understand, and that somehow all of this is just fine. Quantum physics, of course, has unveiled the extreme weirdness of absolutely everything, but the scope of the theory's implications— parallel universes and subjective realities and the ability of a particle to exist everywhere at once—are a lot for us to grok. Even the great religions, with their millennia of wisdom, are more like gateways to unknown journeys than roadmaps of the entire terrain.

Were dolphins, as Ocean claimed, "multidimensional beings"? Were they communicating to us using "holographic images"? Did they "possess consciousness on an entirely different level than humans"? Who knew? A large part of me wanted scientific proof before accepting any such claims. I realized this was a limited way of measuring truth, but it was still the only one that we had. Ocean herself was aware that some of her ideas sounded outlandish; she had a wry sense of humor about this and never imposed her beliefs, or judged anyone else for theirs. Even her son and two daughters (in their forties and fifties now) hadn't entirely signed on, Ocean said: "There's a bit of 'Oh, Mom. Mom does far out things. I wish she'd just be a normal mom.'"

Ocean and I had picked up lunch at a health food store, and I followed her into her kitchen, a homey room lined with knotty pine that opened onto an outdoor deck. The house smelled of woodstove and saltwater, with a dash of cat. It was ideal for Ocean's purposes—she hosts half a dozen workshops each year, and many of the participants stay on the premises. Upstairs there is space for dozens of people to lounge in the living room; downstairs is given over to a spacious meditation room along with multiple bedrooms and communal bathrooms. Two weeks from now, forty-four people were scheduled to arrive for a seminar titled "From Here to Infinity."

As we sat down to eat, Bozzoli walked in. He was a smallish man with a gentle presence. His hair was either brownish gray or grayish brown and it flopped over his face slightly, giving him a rumpled, professorial look. Like Ocean, he had pushed the odometer a bit, but also like her, he could easily be taken for someone much younger. Bozzoli is French, and his accent welled up like music when he spoke. These days, he spent much of his time finishing up a lushly animated 3-D film about the mystical properties of water. "Water has a deep meaning," he said, pulling back a chair and joining us at the table. "And the whales and the dolphins have a lot to teach us about that. Water is like the biggest computer of the universe. It holds the memory of everything, of all that is and that ever was." He fixed me with an intense look. "Water is everywhere. It's in our bodies. You drink it, you pee it—"

"It never goes away," Ocean added. "It's just recycled."

I nodded, eating my soup, unable to add substantively to the conversation. I'd gotten a bit of a sun overdose that morning, and my snorkeling mask had been sucked onto my face so tight that it had given me a whomping headache. I was still processing four hours of swimming with the spinners, and wasn't ready quite yet to discuss the mechanics of the universe. Still, I was interested in what Bozzoli was talking about. Water has secrets. It is the element we fathom the least, and we love it and fear it and take it for granted all in equal measure. "It will take you into altered states, to understand water," Bozzoli said, nodding for emphasis.

These Star Trekian views about dolphins acting as teachers and visionaries and envoys from other dimensions—these were the convictions that had brought Ocean and Bozzoli to Hawaii, and that drew other seekers here too. Just down the hill from Ocean's property, in homes clustered near the water, the residents of Dolphinville were exchanging cell phone calls about their morning swims and reporting on the whereabouts of the dolphins, just as they did on every other day. Earlier, Ocean had told me that she received letters addressed simply to "Dolphinville," the way kids send mail to "Santa Claus, North Pole." She'd also marveled at how, even over the decades, no one who belonged to Dolphinville had died: "Not one person. And anyone who had been sick has gotten well." The implication was that the dolphins had healed them.

"I would rather have truth than mysticism," Ken Norris wrote in a letter dated November 30, 1977, lamenting the emotional ways his study animals affected people. Even back then, it seems, no one wanted dolphins to be ordinary. It was more fun to think of them as extraterrestrials than as the playful fish-hogs of the sea. Ironically, the person Norris was writing to, a neuroscientist named John Cunningham Lilly—who also happened to be Ocean's mentor—was the man most responsible for the trippy, groovy image of dolphins. Lilly was inventive, he was enthusiastic, he had been rocked back on his heels by his studies of the dolphin brain, and he was very insistent about wanting to know: *What if truth and mysticism overlapped?*

CHAPTER 2

BABIES IN THE UNIVERSE

One morning in 1949, a very desirable brain became available in the coastal town of Biddeford, Maine. The brain belonged to a twenty-eight-foot pilot whale that had stranded on the beach during a ferocious Atlantic storm; the animal, which had just expired, lay on its right side half buried in sand, left eye gazing skyward, black body big as a boxcar. Reading news of the whale's demise, Dr. John Lilly, thirty-four, who happened to be in nearby Woods Hole, Massachusetts, at the time, called two of his neuroscience colleagues. "We discussed the possibilities of obtaining the brain of this particular whale, because we wanted to find out whether these animals had very large brains—much larger than ours," Lilly wrote later. Giving the idea a unanimous thumbs-up, the trio gathered hacksaws, axes, and thirty gallons of formaldehyde, and quickly set out on the five-hour drive to southern Maine.

For brain aficionados, this was a bonanza. To date, no one had managed to examine a whale brain in good condition—all of the studies had been done on animals that had been dead too long, their bodies wrecked by decay. Now, suddenly, here was the rare chance to obtain a relatively fresh specimen of what was rumored to be the most compelling brain around.

Upon arrival the scientists scrambled across the beach, and they saw the whale immediately. Actually, the smell preceded the view: "There was a powerful clinging odor," Lilly recalled, "a sort of cross between rotten beef and extremely rancid butter." Soldiering on— "We held our noses frequently"—the three men sawed through blubber and muscle and bone, eventually revealing the huge brain. "We noted that it was very much larger than a human's, more spherical in shape than ours, and looked like two huge boxing gloves," Lilly said. He was struck, too, by the architecture of the brain, its labyrinthine folds and crenellations. By comparison, the cat and monkey brains he'd been studying in his lab seemed puny and dull.

In the end, the brain more or less disintegrated on the beach, having moldered in the hot sun, and Lilly did not get the specimen he'd hoped for. But he left with something more valuable: a lifelong obsession. "I felt awed and diminished," he wrote. "I wondered how such a small hill of flesh lived, what it thought, if it talked to its companions. We were all silenced by the awesome mystery of the whale."

At the time, Lilly's scientific career was under way, filled with promise and breadth. He was a trim man with a hawkish face who had earned a physics degree from Caltech and a medical degree from the University of Pennsylvania; he had researched high altitude aviation for the military, using himself as a crash-test dummy to test the effects of explosive decompression. He was versed in biophysics, chemistry, psychoanalysis, computer science, and neuroanatomy, a commissioned surgeon about to take a prestigious position at the National Institute of Mental Health. It was the brain that fascinated him most, the numinous computer of the mind. To Lilly, the brain was the ultimate black box, an inner space as unknown as the cosmos. It was the doorway to wonder—if we could just get past the threshold. The only way to parse its secrets was to probe the most sophisticated gray matter available, but tinkering with humans was out of the question. The appearance of the pilot whale, with its behemoth brain, raised new possibilities.

A marine biologist friend suggested that Lilly consider bottlenose dolphins—they had similarly high-powered brains, but dolphins were a much handier size to work with. Also, they were available: In 1955, Forrest J. Wood, the director of Marine Studios, a marine park and research lab in Florida, agreed to let Lilly and seven other scientists visit the premises to conduct brain experiments on five bottlenoses.

≈

Every so often, life serves up a fateful cocktail of right person, right time, on the rocks, shaken—with history changed as a result. Dr. John Lilly, with dolphins, in the Cold War fifties and psychedelic sixties, is one such potent recipe. But first—on Lilly's steep, bumpy learning curve—a dozen or so bottlenoses had to die.

To put it mildly, 1955 was a rude time for brain research. Lilly's work included vivisection—surgical experimentation on live cats, dogs, monkeys, pigs, sheep, and rats—and invasive doings with electrical wires. Part of his process involved hammering steel sleeves into the creatures' skulls and then jabbing the exposed viscera to see which areas lit up with pain when stimulated, and which provoked pleasure or seizures or other reactions. Lilly's notes from these experiments contained many sentences like "The animal jumped at every prick of the needle as I injected the fluid, under pressure, into the upper cranium," and "Dr. Mountcastle thrust his arm all the way down through the mouth into the throat. He pulled the larynx out with one finger and inserted a small tube." Using these and other techniques, Lilly and his fellow scientists allotted themselves two weeks at Marine Studios to map the dolphins' cerebral cortex.

In Florida, things did not go well. With dispatch the team killed all five bottlenoses. The first dolphin, injected with Nembutal, quickly lapsed into convulsions and heart failure. The second one did the same, though he was revived temporarily. "We put him back in the tank to see if he could swim or whether his brain had been damaged by a period of anoxemia," Lilly wrote. When Dolphin #2 was placed in the water he listed heavily to starboard and made repetitive, piercing cries. As the scientists watched, the two other dolphins in the tank responded. They swam over and, working together, tilted the injured animal upright and pushed him to the surface. "Immediately a twittering, whistling exchange took place among the three animals," Lilly observed. He was struck by the care the dolphins gave to their tankmate, and he had the impression that they understood the situation and were communicating about it. But Dolphin #2's brain was too ruined to survive, and he was euthanized. The third, fourth, and fifth dolphins followed.

"We were all shocked and saddened at the rate of death under anesthesia," Lilly wrote. "Each death was a new experience to us all." They lacked basic information, the scientists realized, like the fact that dolphins are voluntary breathers: when they lose consciousness, they die; that when they are taken out of the water and subjected to gravity their organs can be crushed under their own weight; that their skin is more sensitive than a human's, abrading painfully or even sloughing off if not treated extra-gently. So the experiment's conclusion was effectively this: not one of the eight celebrated scientists knew a damn thing about dolphins.

It seems unthinkable now, but as recently as the sixties most people had never seen a dolphin. All of the ocean was a tabula rasa, in fact. Jacques Cousteau had just appeared on the scene with his book and documentary *The Silent World*, capturing everyone's attention, but real knowledge was scarce. Dolphins were fish-shaped riddles, roaming the seas largely out of sight and mind and imagination, chasing squid and playing games with jellyfish and strands of kelp, chattering and buzzing and clicking to one another, surfing waves and rubbing pectoral fins in aquatic obscurity.

Lilly was about to change all that.

≈≈≈

The Stanford campus was lit up with autumn when I arrived on a crisp October morning, driving down from San Francisco among the Silicon Valley commuters. The leaves blazed russet and crimson and a zingy, extroverted ochre. Everything smelled like eucalyptus. Students whizzed by on beach cruiser bikes with an intensity more suited to sudden death playoffs than anything resembling cruising, and they sped among the tall shadows of the trees on their way to classes and pitch meetings. The place was stately, but it was hard not to notice that even the air seemed frazzled. I tossed back a double espresso in front of the Green Library and headed into the building.

Inside, across the marble floor and through the Corinthian colonnade, beneath the cool, vaulted ceilings and tucked into the hushed Special Collections reading room, five file boxes awaited me. I had ordered them from an erratic catalog of file boxes that spanned 240 linear feet in total, all the existing "manuscripts, proposals, reports, notes, diagrams, photographs, charts, data, reprints, and extensive audio tape, video tape, and DVD/CD-R material relating to John Lilly's research into dolphin intelligence." Stanford had acquired the archive after Lilly's death in 2001, and it remained mostly uninspected and un-indexed despite widespread curiosity about its contents. Lilly's life was a large one, and he seemed to have hung on to every scrap of paper, every letter and conference brochure and scientific report, every memo and operating manual and flotsam and jetsam receipt, every bit of kaleidoscopic ephemera from his eighty-six years.

"I see you're working with the Lilly files," the reference librarian said to me, sliding over the first gray box. She had close-cropped silver hair and a no-nonsense nautical air. "He was an interesting one." She shook her head. "I'm not sure we've ever quite gotten this down. Can you please let us know what you find?"

Lugging my box to a long wooden reading table, I thought about what that might be. In the decades after Lilly's first, ill-fated dolphin encounter, his career took many hairpin turns. Putting the Marine Studios in the rearview mirror, he resolved to correct his mistakes and continue his research: "I was stimulated and inspired . . . to devote more and more effort, time, and money to these charming creatures." By 1957, Lilly believed he had come up with a safe, anesthetic-free way to access the dolphin's brain. He even experimented on himself, hammering a sleeve into his own skull. Once this was accomplished, it was then possible to insert electrodes and inject chemicals "through small needles anywhere in the brain."

The upside of being able to poke around in there, Lilly believed, was to gain push-button control of a creature's responses. Lilly was fascinated by the sheer size of the dolphin brain—a hallmark, he believed, of advanced intelligence—coupled with the distinctive noises they made, their odd cacophony of sounds. Did these sounds represent their emotional states? Their thoughts and opinions? Bawdy jokes? No one knew. But if you could send a dolphin into spasms of happiness by prodding a particular part of his brain, then in theory you could decode the communications that followed. You could begin to translate his language. Possibly, even, you could deduce what he was doing with that big brain of his.

New technique in hand, Lilly returned to Marine Studios and was presented with Dolphin #6. The skull hammering went smoothly;

almost immediately Lilly managed to locate the pleasure center of the dolphin's brain (and also a spot that made the animal roll his eyes forward and backward). Whenever the electrodes hit this zone the animal would vocalize exuberantly: "Whistles, buzzings, raspings, barks, and Bronx cheerlike noises were emitted." Lilly wondered if the dolphin could be taught to stimulate this area himself as a kind of reward; to test this idea he built a switch that Dolphin #6 could push with his beak. "While I was assembling it," Lilly wrote, "I noticed that the dolphin was closely watching what I was doing."

Dolphin #6 not only learned to reward himself by working the switch, he began to do it before Lilly had even finished the wiring. This instant uptake got Lilly's attention. Other lab animals had required far more time to figure things out: "I had the rather uneasy and eerie feeling that there was a good deal more *purpose* behind this animal's behavior than I had ever seen when working with monkeys." Dolphin #6 pressed the switch so incessantly that it eventually jammed; the dolphin responded with a loud tantrum of every noise he could seemingly make, including, according to Lilly, "an explosive series of air-borne vocalizations." Once the switch was fixed, Dolphin #6 went right back at it, pressing with such alacrity that he pleasured himself into a grand mal seizure. "I suddenly realized that he was going through too intense a stimulation of his brain near the motor cortex," Lilly wrote, adding sadly: "Once again death had come in our experiments because of our ignorance. Slowly but surely we were learning the difference between us and these animals, and how fatal our mistakes could be."

Afterward, Lilly studied the tape of Dolphin #6's outburst. Because dolphin sounds are much faster and higher-pitched than human speech, he slowed the replay down. To his amazement the dolphin seemed to be mimicking phrases used by his human handlers, as well as their laughter. Later, Lilly would present a paper that recounted how Dolphin #6's behavior had affected him:

We began to have feelings which I believe are best described by the word "weirdness." . . . The feeling of weirdness came on us as the sounds of this small whale seemed more and more to be forming words in our language. We felt we were in the presence of Something or Someone, who was on the other side of a transparent barrier which up to this point we hadn't even seen. The dim outlines of a someone began to appear.

This was a pretty seismic shift for a government scientist—and some highly unusual language for the gray-suited fifties. But Lilly wasn't your average scientist. As a boy growing up Catholic in St. Paul, Minnesota, Lilly had visions and premonitions in church; in his early teens, he spent hours debating life and death, the purpose of love, the nature of the universe. At sixteen, he wrote an article for his prep school newspaper posing the question: "How can the mind render itself sufficiently objective to study itself?" At twenty-three, he observed his own mother's brain surgery.

Among the archives, there were at least fifty-three cartons of recordings, from the Marine Studios experiments and many others, of dolphins emoting. On some tapes the animals chittered and squawked to one another, sounds Lilly referred to as "delphinese." On others, human voices can be heard coaching the animals through exercises intended to teach them English, or as Lilly put it, "humanoid." After all, babies started off jabbering and crying, and yet somehow those noises became intelligible speech—so why not dolphins?

~~~

By 1958, Lilly had decided to go all-in, quitting the medical mainstream and moving to St. Thomas in the U.S. Virgin Islands to establish a new human-dolphin laboratory known as the Communication

Research Institute (CRI). His goals for CRI were not modest: "I visualize a project as vast as our present space program, devoting our best minds, our best engineering brains, our vast networks of computer people and material and time on this essentially peaceful mission of interspecies communication, right here on this planet." It would be an entirely new type of facility, a place where researchers would not merely study dolphins but live with them in specially designed buildings. This total immersion, Lilly explained, was imperative if the dolphins were to "learn our language and ways."

The strangest and most bemusing thing about all this, it seemed to me, was that despite Lilly's audacious and quixotic Caribbean plans, his desire to work "continuously at the edge of mystery," and his pronouncements about dolphins being superior to man—every single buttoned-down, tight-lidded government agency you can imagine signed on to the project. Grants rolled in from the National Science Foundation, the National Institute of Mental Health, the Office of Naval Research, the Department of Defense, and NASA, among others.

They all had their reasons for caring about dolphins, most of them quite awful. In the Cold War murk of surveillance and paranoia, Lilly's dolphin brain experiments were proof that minds could be controlled; when the Department of Defense dealt with Russian and Chinese spies, knowing where to stick the electrodes would come in very handy. Over at NASA, they hoped to communicate with intelligent aliens; dolphins seemed promising for a trial run. The Navy wanted to deconstruct dolphin sonar, which was insanely refined compared to a clunky submarine's, and to figure out how the animals made deep dives without decompression.

Lilly had other ideas, too, about how various agencies might benefit from dolphins: "Obviously, if we establish communication it may help us to solve many of our own marine problems." He suggested using the animals to retrieve—and deliver—missiles, pinpoint shipwreck survivors, detect enemy submarines, and patrol the ocean like a finned military police. Operating in a seascape in which they could see but we could not, dolphins would be the ultimate stealth marauders: "In psychological warfare they might sneak up on an enemy submarine sitting on the bottom and shout something into the listening gear . . ." Lilly also mused that dolphins could "help us to obtain new information, data, and natural laws about fisheries, oceanography, marine biology, navigation, linguistics, various sciences of the brain, and space."

It was a lot to ask of an animal, but Lilly set out to create a facility that could contain these goals. "I started from scratch," he wrote, "carving the new facilities directly from the jungle and the wild tropical shore line." There would be a series of seaside laboratories, offices and dolphin pools, connected by breezeways and observation decks. When the island topography didn't cooperate, a Navy demolition team stepped in with explosives and blasted out chunks of the shoreline. Lilly, meanwhile, relocated his family to St. Thomas. In keeping with the theme of mercurial change, he brought a new wife and some new children. Lilly had recently divorced from his first wife, Mary, with whom he'd had two sons, and remarried a fashion model, Elisabeth Bjerg, who had three children of her own. The brood was rounded out to an even half dozen when Bjerg and Lilly had a baby girl right before leaving for the Virgin Islands.

All that remained was to load up the dolphins. In 1960, two more Marine Studios bottlenoses, Lizzie and Baby, were airlifted from Miami. Plans went awry when both dolphins perished shortly after arrival. Lizzie died after being dropped onto a cement floor; Baby succumbed to a bacterial infection. Frustrated, Lilly returned to the mainland, studied more carefully how to transport dolphins by plane, and came back with Elvar and Tolva, a male and female bottlenose.

Almost immediately, Elvar distinguished himself. He was a "bold and pushy dolphin," according to research notes, with an expansive vocal range. Along with the standard whistles and clicks, Elvar emitted "a series of barks, wails, moans, buzzings, trumpetings, banjo-like sounds, and quacking." Despite the eccentricity of the research—one photo I found showed Elvar's English teacher, Ginger Nadell, in the tank with him during a "vocalization lesson," holding the dolphin across her lap—it seemed like progress was being made.

Leafing through photos and documents from the early days of CRI, I was struck by what a halcyon moment it must have been. If you had frozen time in 1961 you would have captured Lilly at the peak of his glory, knee-deep in enviable research beneath the palm trees, surrounded by a bevy of gorgeous female assistants, approaching the border of what seemed like a shimmering new frontier. He also published a book that summer, *Man and Dolphin,* and it began with this bold statement: "Within the next decade or two the human species will establish communication with another species: nonhuman, alien, possibly extraterrestrial, more probably marine; but definitely highly intelligent, perhaps even intellectual." The book was a hit with the public; widespread fascination with Lilly's predictions made him a celebrity. Who *wouldn't* want to converse with a supersmart dolphin? Lilly's circles grew increasingly glamorous: some of the era's most brilliant minds were captivated by CRI's research—Carl Sagan, Aldous Huxley, and Richard Feynman among them. But for Lilly the empirical scientist, it was the beginning of the end.

≈

By 1968, CRI was gone. The eight dolphins in Lilly's program had either died or been released into the sea. Elisabeth Bjerg had left him and taken the kids. All of his funding spent, his staff dispersed, Lilly traveled west to Esalen, a spiritual retreat in Big Sur, California. "I shut my lab because I didn't want to continue to run a concentration camp for my friends, the dolphins," Lilly wrote later, but it's clear there were other factors.

Lilly's problems began when a group of high-profile scientists criticized *Man and Dolphin* for its mix of giddy claims unaccompanied by proof. (If you're going to assert that dolphins can teach us about space, in other words, eventually you'll need to back that up.) "Scientifically unsound and naïve," read the review in *Natural History*. "How not to do scientific research." "Borderline irresponsible . . . Anecdotes are used to launch sweeping speculations," the famed biologist E. O. Wilson wrote, dismissing the book: "Lilly's writing differs from that of Herman Melville and Jules Verne not just in its more modest literary merit but more basically in its humorless and quite unjustified claim to be a valid scientific report." A Navy researcher named Bill Evans summed up the prevailing mood, telling *The Wall Street Journal:* "I find Dr. Lilly's work very interesting. I like Dr. Seuss, too." But bad book reviews alone wouldn't have derailed CRI. Those were merely war drums on the horizon, the distant beat of troubles that hadn't arrived yet. The real damage occurred when Lilly began a run of experiments that involved LSD and interspecies sex.

It's not surprising, of course, that a neuroscientist would dabble with mind-altering chemicals. In the early sixties LSD was a legal substance, considered promising for overcoming trauma and alcoholism, among other afflictions. But Lilly didn't sample LSD so much as fall madly in love with it, immersing himself—and his dolphins—in this new hallucinogenic world. "I read practically everything that had been published about acid and acid trips," he wrote. "I proposed trying LSD on dolphins as an aid to my understanding of the substance and some of the physiological dangers of its use. Each of the six dolphins tested apparently had very good trips with no problems attendant upon their breathing, heart action, or swimming abilities. These experiments gave me confidence to go ahead and try it on myself." (It's hard to know how Lilly determined that the dolphins enjoyed their LSD trips; notes indicate that after being injected with the drug, the animals floated stoically and silently in their tanks.)

Psychedelics, Lilly believed, were the key to gaining access not just to the dolphins' inner lives but to the sanctum of the human brain itself. To experience the heightened states of consciousness he sought, the mind had to be untethered from its usual distractions, floating free in the sea of existence. In his self-experimentation, Lilly took LSD while encased in an isolation tank—a device he had invented for the U.S. Army, which wanted to study the effects of sensory deprivation. The tank was lightless, a dark womb filled with 93-degree saltwater that held a person afloat at neutral buoyancy. Usually the tank was sound-proof as well, but in St. Thomas, Lilly had the dolphins' vocalizations piped in while he was tripping, the better to commune with them.

He also constructed a flooded house, a prototype residence in which dolphins and humans could cohabit. Margaret Howe, a local waitress recruited to the project by Lilly, gamely moved in with Peter, a male bottlenose, for ten weeks, the two of them living in twenty-two inches of saltwater. Each day, Howe would tutor Peter extensively on things like manners, vocabulary, pronunciation, and math.

During the experiment, Howe took many notes. "The first few nights in the flooded room were awful. I was uncomfortable and hardly slept," she wrote. Howe cropped her hair into a quarter-inch brush cut after discovering that she could never really get dry: Peter squirted her when she talked on the phone and splashed her as she was cooking and sprayed water onto her bed. Her groin became chapped. Algae grew on the walls.

Howe described Peter as a "naughty dolphin," prone to ramming and nipping: "I carry a long-handled broom with me for that and ward him off." Though she was convinced that her student's humanoid speech was improving (despite his unfortunate lack of vocal chords), to her frustration he often expressed himself loudly in his native tongue: "I do not respond to his attention-getting clicks and whistles. They mean nothing to me and I make that clear."

Judged by the goals Lilly had intended, the experiment was a fail-ure: "My bed now has three inches of water in it," Howe wrote. "My shins are bruised, up and down, from the constant butting with his nose and the front of his flippers. All of this fatigue was also combined with depression . . . wanting to get away and see some people." She concluded: "To actually live with a dolphin twenty-four hours a day is a very taxing situation."

In another way, Lilly's cohabitation experiment succeeded a little too well. Peter began to follow Howe around the house with a constant erection, begging for sex. *"He does not go away,"* she wrote, italicizing for emphasis. *"This is a problem,* and it must be solved . . . I cannot go on having my shins belted about by lusty little Peter. *It hurts!"* The dolphin, a quick study, soon learned how to pin his roommate into corners. But Howe was nothing if not a trooper, and she eventually convinced herself that the dolphin's attentions were "a very precious sort of thing." At first tentatively and then more enthusiastically, Howe gave in to Peter's desires, reasoning that satisfying the dolphin might create a tighter interspecies bond.

"This is obviously a sexy business," Howe observed. "The mood is very gentle . . . still . . . hushed . . . all movements are slow . . . tone is very quiet . . . only slight murmurings from me." But the frustrations continued: Howe couldn't keep pace with Peter's appetites, despite incorporating hand jobs into his daily routine.

≈

Late in the afternoon, as the sun slanted lower through the library's arched windows, I opened the third file box I'd ordered from the archives. The first item I pulled out was an essay Lilly had written for *Oceans* magazine in May 1977, titled "The Cetacean Brain." The editor had sent Lilly's manuscript back to him in galley form—twenty-one pages grammatically smoothed and typeset for publication—and attached a letter requesting that Lilly approve it with as few changes as possible. I thumbed through the pages: their margins were peppered with notes written in a spidery hand that I recognized as Lilly's. Far from keeping his corrections to a minimum, Lilly had marked up most pages with edits and comments. In particular, he took exception to the copy editor's use of the word "it" when referring to dolphins. For example: "Even if it revives there may be irreversible brain damage leading to death." Lilly had crossed this out wherever it occurred, replacing "it" with "he." Beside these changes he wrote crankily, underscoring with a bold line: "??? The whole point of this article is to remove 'it' from talking about them. Are you or me an 'it'?"

By the time the piece was published, Lilly was back at dolphin research, having spent the better part of a decade engaged in various pharmaceutical and metaphysical pursuits. He had studied gestalt therapy and Gnosticism and est; dabbled in nudism, Rolfing, and hypnosis; spent much consciousness-expanding time in his isolation tank; and discovered an engaging new drug, ketamine, that he liked even better than LSD. The substance, which Lilly called "Vitamin K," was a fast-acting anesthetic that catapulted users out of their bodies for short bursts of time (and unfortunately, as Lilly would learn, it was addictive). He had also remarried, to Antoinetta Lena Ficarotto, known as Toni, whom he met at a party in the Hollywood Hills. "Instantly I knew her and she knew me," Lilly wrote of their first encounter. "We went into a sparkling cosmic love place together." Shortly after that, Lilly and Toni created a new organization called the Human/Dolphin Foundation, based in Malibu.

In one of the files I found a research proposal Lilly had written to the National Science Foundation, dated June 1978. It was an impressive thing, bulky with charts and graphs and diagrams outlining how new computer equipment could be used to "bridge the communication gap" between humans and dolphins. But the effort was futile: by this point, Lilly had been excommunicated from science. His fascination with dolphins and his conviction that their brains held untold secrets was stronger than ever, but his ability to get government grant money was not.

Only a few of his former colleagues stayed in touch, Ken Norris among them. "I think many of your early insights are important, and I don't intend to ignore them," Norris wrote to Lilly, dodging the question of Lilly's later insights, which included statements like "There are times when I feel that each dolphin may be more mentally healthy than the human beings to whom he is exposed." Norris was delicate in his assessment of Lilly's methods: "It's a fine line to draw between imagining and believing. I love to imagine, but belief comes pretty hard with these animals. It's so hard to obtain a decent fact."

"Your letter is a breath of fresh air in the midst of stale cave air," Lilly replied, referring to the attacks from other dolphin researchers. The war drums had grown louder after the publication of Lilly's second book, *The Mind of the Dolphin,* which contained Margaret Howe's notes and an uncensored account of that experiment, and another book titled *The Center of the Cyclone,* about his—and the dolphins'—LSD experiences. It's hard to imagine how Lilly could have expected anything other than wild controversy in the wake of all this, but reading through his correspondence it's clear the antipathy distressed him. One critic, Lilly complained to Norris, had a "vast area of ignorance and a talent for diatribe against matters he cannot possibly understand." Of another, Lilly said, "his obvious lack of experience with dolphins hardly qualifies him as an authority on dolphins." In every line of the letter you can feel Lilly's frustration, and even his pain. "Maybe I take it too personally," he wrote. "I wish it didn't hurt so much."

At the same time, however, Lilly had never been more influential in the public eye. His message that dolphins were unusually clever creatures resonated with people, who were just becoming aware of these animals; his willingness to explore the fringes of consciousness made him a counterculture hero. In 1972, the United States passed the

Marine Mammal Protection Act, a sweeping federal law inspired, at least in part, by Lilly's pronouncements about dolphin and whale intelligence. In 1973, Mike Nichols directed a movie called *The Day of the Dolphin,* a science fiction thriller based on CRI's work, with George C. Scott playing Dr. Jake Terrell, a character modeled after Lilly. Meanwhile, in Malibu that same year, Lilly passed out in his hot tub after a ketamine trip, nearly drowning himself. Throughout 1974, Lilly was taking so many hallucinogens that he had trouble discerning reality from illusion, confusion that landed him briefly in a psychiatric ward and then, later that year, in a short coma. His book *Simulations of God: The Science of Belief* came out in 1977; in it he described himself as an "extraterrestrial who has come to this planet Earth to inhabit a human body." Unfortunately, he added, "the vehicle is too small to contain the passenger." In 1979, the movie *Altered States,* based on Lilly's adventures in his isolation tank, went into production with William Hurt starring. It was an interesting decade.

Also in the seventies, biologist Roger Payne introduced the world to the astonishing songs of humpback whales: intricate, keening calls that echoed through the oceans, containing messages we could only dimly guess at. Up in British Columbia, another scientist, Paul Spong, was investigating orcas, and had come to the same conclusions as Lilly: "My work now is concerned with trying to reveal—or even just glimpse—a new kind of intelligence, something humans aren't even aware of." Spong's study subjects loved music (particularly flute and violin), lived in tight family groups, and possessed a number of startling traits, including a sense of humor. Like their bottlenose and humpback brethren, Spong discovered, orcas communicated with a wide repertoire of sounds: "echolocation clicks, burst-tones, pure tones, whistles, hornblows, frequency-modulated screams, and sounds that I cannot really find any human words for."

As our understanding of the ocean grew, the idea of it as a wondrous parallel universe came along too. Anything was possible: we knew so little. We didn't know, for instance, that octopi had long-term memories, or that great white sharks could be a little shy. We had no idea that penguins could *fly* underwater at twenty-five miles per hour. No one knew much about the lives of cantankerous pilot whales or extreme-diving belugas or any of their kin. Dolphins specialized in mystery, and Lilly was an enthusiastic explorer in the realm of the unknown. It was the place he loved most, in fact.

It was appropriate, therefore, that Lilly's return to dolphin research was funded by curious individuals rather than government institutions, and that the Human/Dolphin Foundation's board included celebrities like Jeff Bridges and John Denver. Lilly's aim this time around was pretty much the same as it had been last time: to talk to dolphins. But now, in 1979, technology had advanced. Lilly planned to use the fast new Apple II computer to generate a third language that both humans and dolphins could understand, a code based on the dolphins' own clicks and whistles. He named this project Janus, after the two-faced Roman god of gateways. (It was also the less romantic acronym for "Joint Analog Numerical Understanding System.") A private patron donated $10,000 to get Janus rolling; Lilly used it to buy two bottle-nose dolphins, Joe and Rosie.

In general, budgets were tight. In file after file I read evidence of financial angst: people complaining about lack of reimbursements from petty cash, scuffles over sums as small as $10.89, voluminous correspondence about how to coax General Motors into donating a van. Writing to one prospective donor, Lilly laid it out straight: "For $7,000/month for 5 months we can obtain the critical answers to the major question: are we able to get a dolphin to learn/select a computer useable code?"

I wish I could tell you that this story had a perfect ending, with Lilly finding his grail and dolphins revealing all through a miraculous interface powered by the Apple II. Instead, the project dissolved in an acid bath of staff squabbling, cash shortages, vicious jealousies, and all-around dysfunction. There was a freewheeling quality to Janus's operations, perhaps because the dolphins and staff were located in Redwood City, up near San Francisco, while Lilly and Toni resided in Malibu. Reading through old memos and records, I had the impression of a high-school class running amok when its teacher left the room. I came across meeting minutes that reported: "The issue of sexual relationships among staff members was discussed at some length, including specification of potential advantages . . ." Someone was accused of "spending too much time on personal vendettas." One manager wrote a memo that said, "I have been harsh, at times picky, and generally merciless in this mud-slinging. And I own that." Documents indicated there had been a staff walkout over pay, and a full mutiny against a particular engineer. Project Janus also included many volunteers, one of whom wrote a letter to Lilly complaining that, "while we have been quarrelling and mucking about in the petty mire of human egos and self motivations, the realization of someday communicating with another species on our planet remains no closer a reality . . . perhaps it is best that Cetaceans never come to fully trust man."

For three years, Janus's software and hardware was fiddled with and tweaked, code rewritten and machines recalibrated, but the dolphins had not broken through. Joe and Rosie had learned only a few phrases; the one word they both picked up unequivocally was "fish."

The burned-out Lillys retreated to Esalen to consider a plan. I was riveted by the transcript of their visit, during which they consulted with a woman named Jenny O'Connor, a channeler of entities known as "the Nine," who introduced themselves as extraterrestrials from the star Sirius B, here to help guide mankind. "I'm totally worn out. Totally worn out," Lilly told them. "Because I've been trundling around raising money, and being the scientist in doing the work the way I think it ought to be done, and fighting with other scientists who think it ought to be done in another way, and all this shit." The project, he complained, was "getting dull." Toni agreed: "My fantasy is . . . of being in the tropics, with warm water, being able to play with the dolphins, just any way—I don't care how scientific it is."

Through O'Connor, the Nine recommended that Lilly simply disappear from view. His legend would be assured, they said, if he suddenly went AWOL, rather than wrangling with Janus any longer: "Mankind likes the hero that rides off into the sunset."

Perhaps Lilly considered this, but what he did instead was trudge along with the project while Toni tried to relocate it to more idyllic surroundings. The file box also contained a mock-up brochure for a new "interspecies resort" in Costa Careyes, Mexico, near Puerto Vallarta. The development would include a hotel and spa, a research facility, and a bay in which people and dolphins could "ultimately meld their relationship." The dolphins themselves would be free to come and go, with no enticements except for live music, played from a floating raft. Guests would be issued a backpack-like device called a "Dolphin Talker" so they could communicate with wild dolphins "as soon as it has been established that this is feasible." Janus would carry on its work with Joe and Rosie south of the border.

Reading through this material, I was struck by Lilly's persistence. Despite the fallout from CRI, the unappealing experiences of Margaret Howe, and the inertia of Janus, the man truly, deeply believed that humans and dolphins belonged together, and that we could learn to speak one another's language. Though Lilly's Mexican dolphin resort would not come to be, his vision, however embattled or far-fetched, remained.

≈≈

Lilly lived to see the new millennium, but Toni did not. She died of bone cancer in 1986. Neither years nor heartbreak made Lilly any softer about the human race; if anything, his opinion of mankind grew gloomier. "If I were from an older, greatly superior culture from some other place in the vast universe, I would recommend that this planet be shunned," he wrote. "The human species is so arrogant that it doesn't recognize its own superiors. The only way that humans in the mass will respect any other species, apparently, is the ability to beat them in warfare."

From what I'd read of Lilly's writings, from what I knew of his life, I suspected that he was drawn to dolphins for the same reason he liked to take drugs and retreat into his isolation tank: as an escape from human society. Working as a scientist during World War II, the Vietnam War, and the Cold War, Lilly had a front row seat to the havoc people can get up to, the low road we often seem to travel: "The whole philosophy that says that the one species must rule the other species has been cast out of the thinking of myself and my colleagues," he reflected. "We are often asked, 'If the dolphins are so intelligent, why aren't they ruling the world?' My very considered answer is this—they may be too wise to try to rule the world."

Although he couldn't prove them, Lilly's ideas about dolphin communication were hardly dead ends. Since his initial explorations, other scientists have had great success barking up this same tree. In the Bahamas, biologist Denise Herzing, founder of the Wild Dolphin Project, has created a portable underwater device that sounds a lot like Lilly's Dolphin Talker, except that Herzing's seems to work. She uses pattern recognition software to translate dolphin sounds into English, swimming with a pod of spotted dolphins on the Bahama Banks. Recently, science journals reported that one of Herzing's study dolphins had whistled the code for "sargassum," a type of seaweed the researchers had been using as a toy.

During the last decade of his life Lilly lived on Maui, surrounded by the inhabitants of what he called the "Cetacean Nation." The spinners, the spotteds, the bottlenoses, the orcas, the humpbacks, every species of dolphin and whale—they were all tribes that belonged to the same huge clan, one that Lilly believed deserved at least some of the same rights we had. He became an antiwhaling advocate, attending International Whaling Commission meetings and traveling often to Japan. As an old man, Lilly had the restive eyes of his younger days, and a mane of white hair. He would often sit for hours and stare out to sea.

It was as though Lilly knew he would be judged, and where his priorities lay: "A certain willingness to face censure, to be a maverick, to question one's beliefs, to revise them, are obviously necessary," he wrote. "But what is not obvious is how to prepare one's own mind to receive the transmissions from the far side of the protective transparent wall separating each of us from the dark gulf of the unknown. Maybe we must realize that we are still babies in the universe, taking steps never before taken. Sometimes we reach out from our aloneness for someone else who may or may not exist. But at least we reach out, and it is gratifying to see our dolphins reach also, however primitively. They reach toward those of us who are willing to reach toward them."

CHAPTER 3

CLAPPY, CLAPPY!

If you wanted to swim with a dolphin at Ocean World Adventure Park, a facility billed by its owners as "the most advanced marine interaction park of its kind," you would first have to travel to the Dominican Republic, a Caribbean country about twice the size of New Hampshire. Though it adjoins poverty-stricken, earthquake-ravaged Haiti, the Dominican Republic is a lively place, buzzing with movement and merengue music. Driving west from the Puerto Plata airport is like being plunged into a pinball machine, motorbikes and lorries and mini-cars streaking past hot pink casitas and tangerine bodegas and canary-yellow tiendas in a blur. After a few miles of crowds and chaos the road mellows out and the tourist suburbs emerge, with their cheap stucco and bougainvillea. They spill along the coast in clumps of time-shares that all look the same, but one landmark is hard to miss: the thirty-foot cement-and-neon dolphin at Ocean World's entrance.

Inside, beyond a fortress of gates and past the diesel fumes of a bus-unloading zone, accompanied always by jaunty music played at startling volume, you would face decisions. You would have to decide if you wanted to purchase the Royal Dolphin Swim ($199 for sixty minutes in the water with dolphins), or the Dolphin Swim ($169 for thirty minutes), or perhaps the Dolphin Encounter (a $109 wading opportunity for nonswimmers), or if you were feeling flush, the immersive $250 Trainer for a Day experience. Regardless of which of these options you selected, you would then need to determine if you wanted to add on a sea lion encounter ($79), or a shark and stingray encounter ($79), or perhaps both. That is, unless you were pregnant, in which case you may not do any of the above.

I stood in the ticket line, looking at mounted posters of people posing with the creatures that lived inside these gates, including a pair of Bengal tigers in a faux stone grotto, caged behind a sheet of plexiglass. (In photographs the plexiglass is invisible, creating the illusion that the visitors are bravely facing off with the tigers.) The photo nearest to me featured a woman in a bikini gripping a dolphin by his pectoral fins; the end of the dolphin's beak was scraped off and raw pink. In another picture, a grinning man in a lifejacket kneeled in the water with a dolphin on either side of him, their beaks nuzzling his cheeks.

Ocean World is a place that encourages touching. At other marine parks, patrons must be content to sit in the bleachers and watch dolphins doing choreographed tricks; here they can pay to become part of the show. "The dolphins are trained to interact with each person regardless of his or her size, shape, gender, or abilities," the park's Web site reassures. Depending on which packages they purchase, visitors might ride a dolphin by latching onto its dorsal fin, or be kissed on the mouth by a dolphin, or have a dolphin pretend to dance with them. They might be licked on the head by a sea lion, or pet a stingray with its stinger removed, or stroke a nurse shark's sandpapery skin. For anyone uninterested in that, there is also a dolphin-themed casino and discotheque overlooking a marina with its own customs-and-immigration facility.

On the day I stopped by, it was the doldrums of winter and I expected to see flocks of sunshine-seeking tourists streaming through Ocean World's gates, but the place was not crowded. I walked in and paused for a moment to get my bearings: Ocean World occupies a big swath of seafront. There was a brimming gift store situated directly in front of the entrance and exit, across from a kiosk where people could buy photos and videos of themselves with the animals, taken by Ocean World photographers. At the front of the gift store a display overflowed with stuffed dolphins and tigers.

The last time I had visited a marine park was in the late 1960s, in Fort Lauderdale, Florida, when I was four or five years old. Coincidentally, its name had also been Ocean World. In my misty childhood memory it was an exotic place, filled with creatures I'd never imagined, like hammerhead sharks, alligators, and—I was from Canada—parrots. There were dolphins there too, and when they weren't performing their dolphin show the public could fondle and feed them, dropping minnowy little fish into their mouths. Though it was an artifact of an unslick era, Ocean World Fort Lauderdale kept its doors open until 1994. It finally closed after being charged with animal abuse by the U.S. Department of Agriculture. Among other violations, it was cited for cramming its dolphins into tiny overchlorinated pools, inadequate veterinary care, dumping sewage into the Seminole River, and "the improper burial of dolphins in the park's landscaping." One time, management had attempted to paint the inside of the dolphin tank without first removing the dolphins. The park also lost a lawsuit brought by a visitor, Ernest Coralluzzo from the Bronx, New York, who was bitten by a bottlenose while standing next to the petting pool. Coralluzzo charged that the dolphin, Dimples, lunged at him without provocation, causing nerve damage to his left arm. During the trial, it came out that Dimples was a repeat offender, having bitten at least six other people and walloped several more with his tail. Despite having treated the dolphin abysmally, the president of Ocean World defended him in the press: "It was an accident . . . Dimples is part of our family."

Now, decades later, I had come to this Ocean World to find out what life was like for a dolphin inside "the largest man-made dolphin enclosure in the world," because it is impossible to talk about dolphins without considering this fact: people will pay a lot of money to be with them. What other animal could entice a family of four to drop $800 in one afternoon? Certainly, no one here was spending $199 to kiss a stingray, or buying $120 photo packages on account of the toucans or the sea lions or even the plexiglass-encased tigers. It was only the dolphins who commanded these fees.

The weather in the Dominican Republic was searingly hot, and the unshaded walkways and dolphin pools baked in the tropical glare. I stopped by the sea lion tank and took a long pull from my water bottle. Much concrete had been poured here, contributing to the fiendish temperatures. Beneath the peppy samba music, Ocean World had a vaguely militaristic feel, with armed security guards posted around the property. NO GUNS OR WEAPONS ALLOWED WITHIN OCEAN WORLD PREMISES, warned a sign on the side of a building.

The park's sixteen dolphins were housed in a lagoon divided by a lattice of pathways and docks. There were one or two dolphins in each pen; below the docks, wire fencing separated the enclosures. Most of the dolphins hung at the surface, barely moving, a resting behavior scientists refer to as "logging." One of the dolphins trailed alongside me, staring up while I walked. Another nudged halfheartedly at a floating red ball.

Every dolphin here was a bottlenose, which was not surprising. Marine parks rely on this species because they tend to fare better in captivity than other types of dolphins, and learn quickly to perform the requisite tail-walking moves or synchronized leaps. (Orcas and beluga whales are popular attractions too, though their size makes them more difficult to keep; when removed from the wild, both species die at alarming rates.) Occasionally, the parks have also exhibited less familiar species like Amazon river dolphins, Risso's dolphins, pilot whales, white-sided dolphins, melon-headed whales, spinner dolphins, Fraser's dolphins, Commerson's dolphins, pantropical spotted dolphins, striped dolphins, common dolphins, and false killer whales—but none of them lasted too long. History has shown that dolphins of every species, bottlenose included, live fretfully in tanks, and though some individuals adapt better than others, many fall short—and often far short—of their natural life expectancies.

I knew this fact and I found it upsetting, and I'm not a big fan of canned tourist experiences, especially when they are as transparently craven as this one seemed to be. As a rule I don't enjoy watching tigers caged behind a casino, or dolphins conscripted to tow kids on boogie boards around a fake lagoon. At the same time, I wanted to find out what was going on in these tanks, if there was any possible benefit for humans or dolphins in this arrangement. I am a fan of top-notch facilities like the Monterey Bay Aquarium in Northern California, where every exhibit is rooted in the latest marine science, and where admission profits are funneled right back into important ocean research—but trained dolphins are not part of their mandate. Dolphin shows and dolphin petting zoos and big-ticket dolphin encounters are another business entirely, probably because each new finding about dolphins reveals them to be even more astute and self-aware and socially advanced than we previously realized, and this in turn calls into question the ethics, and the research benefits, of keeping such sophisticated creatures penned up, making them do our bidding in order to eat. Science, in other words, does not coexist easily with the Royal Dolphin Swim.

But people don't always concern themselves with the details, and they come to marine parks like this one and bring their children and plan vacations around dolphin captivity—and I would imagine that these people are well intentioned and would insist that they absolutely cherish dolphins, even as the animals orbit their teacup pools as endlessly as lost satellites. Relentless public relations' efforts support this schism, and the idea of dolphins as happy ambassadors to all humanity. To hear the marine parks tell it, they are providing a vital educational service, one that aids dolphins and engenders love for the ocean and helps ensure the long-range conservation of the animals that live there. All of which are worthy endeavors, of course, and I was prepared to upgrade my impressions if any of these claims were true.

I had opted for the thirty-minute dolphin swim, so I changed into my bathing suit and followed the signs to a snack bar called the Dolphin Hut. There, I was issued a yellow lifejacket and herded in with a group of seven other people. We bunched together in a patch of shade while an Ocean World employee laid out the rules, holding up a stuffed dolphin for demonstration purposes. "Do not hit them or smack them," he said in Spanish-accented English. "They don't like that. And don't touch them in the blowhole. That is their private area."

"We just did the shark touch," a chipper blond woman told me. She pointed to her husband, a lanky man with sparse hair who was sweating profusely. "He went in the tank. I wouldn't."

"They're just nurse sharks," the man said, looking exasperated. "They don't even have teeth."

The woman shook her head. "I don't care. They're *sharks*."

Orientation over, we were directed down the dock to a semicircular enclosure, where two trainers awaited. The water in this pool was darker blue and intended, perhaps, to evoke an actual pond or miniature lake, with little visibility below. One of the trainers stepped onto a floating platform and set down a plastic cooler. The moment he did this, two bottlenose heads popped up in front of him. They had the excited, expectant vibe of Labrador retrievers, except they were twelve-foot-long, 1,000-pound wild animals, which became apparent when I saw them up close. These dolphins made the spinners in Hawaii look like bath toys.

We sat in a row on the side of the platform, our legs dangling in the water. The dolphins dove, shot around the tank, and then reappeared in front of the cooler. The second trainer, a sturdy Latin guy I'll call Alonso, began by introducing them. "This is Serena and Niagara," he said, pointing to the dolphins. They had their heads out of the water and their mouths open and I could see their thick pink tongues, lined by perfectly spaced teeth that looked like sharpened pine nuts. Both dolphins had silvery bodies and the softest, palest pink tinge on their underbellies. "They like being together," Alonso told the group, adding: "If they don't get along, they fight." He made boxing motions with his fists. The dolphins stared hungrily.

Alonso didn't seem like a bad guy. He just seemed like a guy who would have been happier behind the bar of a Carnival cruise ship, slinging daiquiris on all-you-can-drink night. What he was not, most obviously, was someone who would "exchange and disseminate current knowledge, research and other information" about dolphins, as the International Marine Animal Trainers Association's Web site states as a mission. Opening the cooler, Alonso flicked two teeny translucent fish at Serena and Niagara. "Clappy, clappy!" he yelled, waving his hands spasmodically. The two dolphins rose out of the water vertically and beat their fins on the surface, soaking the people, who screamed in delight.

During our allotted thirty minutes the dolphins performed a sequence of tricks, none of which seemed to engage them even slightly, and all of which seemed, if you stopped for half a second to think about it, depressingly dumb. Serena and Niagara vaulted over our heads in tandem, and dragged us around the tank, and pressed their beaks to our faces on command. Whenever they completed a trick, they raced over to the cooler and resumed their begging posture. Alonso barked instructions at the people; the dolphins responded to hand signs and shrill whistles.

In the end, not even the concrete gloom of Ocean World could squelch the magic of Serena and Niagara, who seemed in all ways more interesting than their trainers, and who were obviously capable of feats on a completely different scale. As far as educational experiences go, however, this one had about as much substance as a dolphin-themed ride at an amusement park. The animals' natural history, so rich and transfixing, was not part of the program. Not a whiff of an ocean conservation message was offered. Even if you ignored the photographer sharing the platform with the trainers, Ocean World's intentions could not have been clearer—and they didn't involve illuminating the incredible real lives of wild dolphins for anybody.

As Serena and Niagara completed their routine, I was struck by the glazed-over look in their eyes and the intensity of their focus on the fish, as though they had been waiting for that cooler to appear for a very long time. While I was thinking about this, I heard Alonso ask the group: "Do you know how long these dolphins live?"

"Twenty years?" a teenage girl answered, guessing.

"Yes!" Alonso said, grinning broadly and revealing a set of mismatched teeth. "But only in here. In the ocean they live shorter." He held up a cautionary index finger. "In the ocean their life is very hard."

≈

It was P. T. Barnum, the circus showman, who first exhibited dolphins back in 1861. Upon discovering that vast beluga pods congregated in northeastern Canada, he collected two of the animals and shipped them to his American Museum in New York. With their goofy Casper the Friendly Ghost appearance, their impressive size, and their sweet dispositions, the belugas were a huge draw in the hours they survived. They shared the second floor of the museum with a taxidermied elephant, an electric eel, a seal who played the harmonica, a fat lady, a one-armed, one-legged soldier, and a giantess named Miss Swan, who sat in an oversize, thronelike chair.

"In August last I succeeded in bringing to the Museum two living white whales from Labrador," Barnum wrote. "One died the first day and the other the second day. Even in this brief period, thousands availed themselves of the opportunity of witnessing this rare sight. Since August I have brought two more whales to New York, at enormous expense, but both died before I could get them into the Museum." Undeterred, Barnum returned to Canada to snatch up more belugas from a seemingly limitless supply, their commercial potential diminished only by the troublesome logistics of carting around an eighteen-foot-long, 3,000-pound animal in a tub of saltwater. At least nine belugas rotated through Barnum's exhibit, all of whom died promptly. Then, in 1865, the museum burned to the ground, taking another two belugas with it. In a lengthy article about the fire, *The New York Times* mourned the loss of the whales and their "fearful death by roasting."

Other enterprises took to exhibiting belugas—the animals were easy to catch—including the New York Aquarium on Coney Island, at the time run by a circus manager and an animal dealer; the Boston Aquarial Gardens; and the Royal Aquarium in London, which was able to keep its beluga alive for four days. Barnum continued as well, building a new museum and restocking it with whales until a fire consumed it, too, along with yet another pair of belugas. In 1897, the New York Aquarium relocated to Battery Park and hauled in two fresh belugas. One died in less than a week. The other beluga lasted for a heroic twenty days before choking to death on his food.

Bottlenose dolphins leaped onto the scene in 1938, with the opening of Marine Studios in St. Augustine, Florida. Billed as the world's first "oceanarium," Marine Studios was originally built as an underwater movie set, stocked with photogenic marine creatures who could be filmed through two hundred portholes—situated for optimal camera angles—in two half-million-gallon steel tanks that hunkered on the beach like twin spaceships decorated in snappy Art Deco style. It was a grand project, optimistic and expensive; someone came up with the idea of selling tickets to the public to offset operating costs.

From the start, the place was a hit. Thirty thousand people showed up on opening day to peer through the portholes at the gleaming coral reef, the schools of jewel-colored fish, the sea turtles and moray eels and manta rays and tiger sharks gliding by. In 1938, America was limping out of the Great Depression, scuba had yet to be invented, and no one had ever seen such a vista before. A helmeted diver walked the tank floor feeding the barracudas and tarpon and goliath groupers by hand. Sun-dappled seahorses tipped in the current; moon-colored octopi jetted under rocks. It must have been hypnotic and wonderful, like a submarine dream of the ocean, now available on land.

Marine Studios' creators included Cornelius Vanderbilt Whitney, heir to the Vanderbilt and Whitney fortunes and one of the founders of Pan Am Airlines; his cousin William Douglas Burden; and Count Ilya Tolstoy, Leo Tolstoy's grandson. The men had ties to Hollywood, and also to the American Museum of Natural History. They were explorers and adventurers themselves, and alongside the oceanarium they built a research lab where scientists could come to study the sea's unknown inhabitants. (Demand for the lab was so high that it was booked a year in advance; John Lilly's first dolphin experiments took place there.) Movie stars and beauty queens were frequent guests, posing for publicity photos. Ernest Hemingway arrived to drink at Moby Dick's, the onsite bar that rocked like a ship at sea.

Marine Studios' most crowd-pleasing feature was its lone dolphin, who swam around carrying a sign in his mouth that said I AM A BOTTLENOSED DOLPHIN. This wasn't the bottlenose's premier public appearance—the New York Aquarium had some as early as 1913—but it was here that dolphins were first seriously trained to do tricks. To everyone's enthusiasm new animals were constantly added to the tanks, including more bottlenoses and spotted dolphins and, in 1948, four pilot whales, the survivors of a forty-six-whale stranding that occurred on the beach in front of the oceanarium.

The four whales were scorched with sunburn by the time they were transferred to the tank, their skin a mess of blisters. They huddled together, even at night, with their bodies always touching. Three of the pilot whales died within eight days. The fourth, who had been named Herman, survived for nine months. Herman was a small pilot whale, probably a yearling, and after he lost his remaining podmates he was bullied by a gang of three male bottlenose dolphins. The bottlenoses were remorseless. The whale, for reasons of fear or temperament or inability, never fought back. Like all cetaceans, Herman was highly vocal, and his sounds were unusually plaintive. An observer described them as "the peevish whining of a young child," and "the crying of young porcupines or beavers." The whale had good reason to whimper. Herman was bitten and rammed and chased. His ribs were cracked. His tail was bruised so badly that he couldn't use it properly and had to swim with a side-to-side motion, like a shark. One time, a bottlenose smashed into the whale with such ferocity that his body was ejected from the water. By the time the staff intervened, removing the assailants, Herman had suffered a broken jaw. He died soon after that.

Dolphins came and went during the fifties, plucked from nearby waters by Marine Studios' forty-eight-foot collecting boat, *The Porpoise*. In the continuum, it became clear that some animals were smarter and easier to manage than others. "Periods of accomplishment would be followed by periods of remission and bad temper," read one account of Marine Studios' dolphin training efforts. Eventually, a star emerged: a gregarious male bottlenose named Flippy. Advertised as "The World's Most Educated Dolphin," Flippy could shoot basketballs and throw footballs and push a dog around on a surfboard. He could fly through hoops. In 1955, he starred in *Revenge of the Creature*, the sequel to the movie *Creature from the Black Lagoon*.

Flippy was soon eclipsed by Flipper, the charismatic bottlenose film and television star. Though Flipper was a male character, he was played by a rotating cast of five female dolphins, all of seemingly invincible talent. After the *Flipper* movie grossed $8 million in 1963, the dolphin, a kind of aquatic house pet on steroids, was given his own TV show. Each week, Flipper rescued his adopted human family from oceanic perils such as underwater explosions, crocodile maulings, and shark attacks. The show's plots were cartoonish and fantastical but they struck a booming chord. "The dolphin does amazing things," *The New York Times* enthused in its review. "It registers pathos and joy in its own way, and manages to upstage anything less than eight-feet long."

Flipper's popularity tsunamied out from Coconut Grove, Florida, where the show was produced, to the rest of North America, splashing into even the most landlocked of places. Suddenly, everybody wanted to see dolphins in action. In the dolphin-mania that ensued—which has only heated up since Flipper's time—marine parks have grown into a multibillion-dollar global industry. Swim-with-dolphins programs are now being launched at the rate of two per year throughout the Caribbean, and can be found in such unlikely spots as Romania and Cambodia.

Marine Studios, fallen from glamor and renamed Marineland Dolphin Adventure, is now owned by the Georgia Aquarium. For $99, its patrons can have a fifteen-minute encounter with a "dolphin artist" who will paint a picture for them, brandishing tubes of paint with his beak. (Trainer for a Day is available there, too, for $475, making Ocean World look like a comparative bargain.) In at least one instance, the dolphins even have shareholders: SeaWorld Entertainment Inc., the biggest brand in marine parks, went public on the New York Stock Exchange in April 2013, with a market capitalization of $2.5 billion. "Growing attendance is not our focus," SeaWorld's CEO, Jim Atchison, told *The Wall Street Journal* after the company's IPO raised $702 million. "Our focus is driving our financial performance."

≈

The enclosure emptied of people, and the trainers packed up their cooler and wandered off too. I stood on the walkway watching Serena and Niagara circling on autopilot, visibly aware that the fish were gone. Electronica pounded through speakers—it was inescapable on Ocean World's grounds—loud enough to make the railings vibrate. I wondered what it was like for such acoustically sensitive animals to be surrounded by earsplitting music and shrieking people all day. Sound is a physical force; when we're bombarded by it we come apart quickly (imagine being locked in a room with someone running a jackhammer or belting out karaoke or scraping over and over on metal). Excessive noise bugs us and hurts us; it damages nervous systems, circulation, mental health, ears. Dolphins, with their ability to hear across far wider frequencies, are especially vulnerable. After a sixteen-hour rave was held near their enclosure at Connyland, a Swiss marine park, two dolphins died.

In the annals of bad dolphin husbandry, Connyland had long been infamous. Previously, its owners built an underwater nightclub with windows that flashed lights and reverberated music directly into the dolphin pool, a feature described by one scientist as "a perversion of the highest degree." In Connyland's thirty-year history its dolphins had also suffered from skin lesions, pneumonia, brain damage, heart deformation, kidney problems, and mushrooms lodged in their intestines.

It's unclear how a dolphin ends up jammed with mushrooms, but dolphins are inquisitive about objects, and will often pick up and swallow things that get dropped in their enclosures. Scientists have puzzled over why this happens so frequently in captivity, why dolphins with their fine-tuned sonar might suddenly mistake a leather glove or a french-fry container for food. They suspect that when marine parks train dolphins to eat dead fish—in the ocean, of course, they hunt live prey—the animals become confused and start nibbling at anything they encounter. This is a dangerous practice for dolphins. It causes intestinal blockages, which are usually fatal. Pieces of plastic are a common killer, but dolphins have also died from ingesting bottle caps, coins, car keys, coffee cups, roofing tiles, cigarette lighters, balloons, rubber toys, jewelry, steel wool, nails, and chunks of asphalt. (A few years ago in Fushun, China, two dolphins ate strips of their tank's vinyl lining and were saved by Bao Xishun, a 7'9" Mongolian herdsman who appears in the *Guinness Book of World Records* as "The World's Tallest Man." When surgical tools failed, Xishun reached down the dolphins' throats with his forty-two-inch arms and extracted the plastic.)

The other likely culprit is boredom. If your entire world were a featureless swimming pool, you would probably take great interest in examining the stuff people tossed in too. For a creature as smart and creative and sociable as a dolphin, there's an undeniable padded-cell quality to life in a blank white concrete tank. Behavioral biologist Toni Frohoff, who studies stress in captive dolphins, has watched the animals gnaw on cement enclosures until their teeth are ground down, and bang themselves repeatedly against the walls. Stressed-out dolphins, like stressed-out humans, develop ulcers, heart trouble, blown-out immune systems. They succumb easily to illnesses like pneumonia and hepatitis and meningitis.

Some dolphins learn to tolerate captivity, but even the most spacious enclosures can never match the open water, where the animals range far in close-knit groups, hunting and playing and socializing in ever-changing conditions. Out there, in the depths and the shallows, they work as a team, devising ingenious fish-catching strategies on the fly. In Shark Bay, Australia, one group of bottlenoses uses sea sponges like protective gloves, covering the tips of their beaks with them when they forage in the rough sand on the seafloor, digging for burrowing fish. In other locations, dolphins blow bubble curtains or stir up mud plumes to entrap schools of fish. Orcas have been filmed lining up in a precise row and using their tails to generate waves that flush seals off an ice floe. In every environment, the dolphins will figure out a plan. In heavy surf, in calm bays, in gin-clear tropical seas and sediment-thick rivers, in kelp forests and eelgrass channels, in moonlight and sunlight and light swallowed whole by the depths, in the mercurial, bountiful oceans, they go about their business together.

Only now are we beginning to understand what that word, *together*, means to a dolphin. In the wild, dolphins are minglers, gadabouts, flirts. Their existence revolves around relationships. Like us, dolphins form intense, long-term attachments with others and maintain them over time, even when separated for extended periods. Scientist Jason Bruck, from the University of Chicago, proved that dolphins recognize their friends' signature whistles even after twenty years apart, and react with obvious excitement when they hear them. Their bonds are so strong, in fact, that when dolphins are in jeopardy they will not leave one another even if it costs them their lives. And when dolphins do lose a loved one, they behave in ways that suggest deep grief.

Often, marine parks tout captivity as a luxurious setup for dolphins because it saves them the trouble of finding their own fish in our increasingly ruined oceans. "Be wild, be free, is just not a valid premise any longer," Tom Otten, the former director of the Point Defiance Zoo and Aquarium, declared in a newspaper interview, summing up that line of reasoning. But even if the dolphins are being served tuna sashimi in suites at the Four Seasons, the truth remains that in captivity their lives are hollowed out. In a concrete tank none of their expertise is needed, none of their exquisite adaptations developed through eons in the ocean, not their sonar or their hunting skills or their communication abilities. The pods that sustain them are no longer part of their world. Instead, their social groups are determined by marine park staff. Their behaviors are dictated by what audiences want to see.

No wonder dolphins get sick, neurotic, depressed, and anxious. No wonder marine parks mix antibiotics and Valium and Tagamet into their food. Like us, dolphins express their emotional states in personal ways. Some dolphins mope and despair. Others become fixated on sex, trying to mate with dolphins of the opposite sex, the same sex, their trainers, and even inanimate objects. When it comes to partners, they are willing to experiment: Marine Studios had one dolphin that continually tried to mount an eel.

They also vent their feelings through aggression. As Alonso succinctly put it, "They fight." In a 2008 issue of *Soundings,* the International Marine Animal Trainers' Association magazine, Steve Hearn, head of the "dolphin department" at Dolfinarium Harderwijk in the Netherlands, described the process of introducing young dolphins into an existing group of male bottlenose adults: "This has to be one of the most exciting and fragile training and socialization procedures we have taken on, because (to be honest) we have absolutely no idea how the dolphins will react to one another . . . Eventually, there is no way around it, at some point you are going to put these dolphins together." Though Hearn and his trainers may be in the dark, as Herman the pilot whale discovered, this type of dolphin mixer often ends badly.

The list of dolphin-on-dolphin injuries and fatalities is impossible to tally—every marine park contends with this and none are too keen to publicize it—but you do not have to look very hard to turn up brutal accounts. Tooth raking, jaw clapping, tail lashing, head butting, biting, and high-speed chasing are common behaviors among agitated dolphins; animals have died from skull fractures after leaping out of their tanks to escape the beatings. In China, a bottlenose had her dorsal fin amputated after what the Dalian Laohutan Ocean Park referred to as "internal strife." At SeaWorld San Diego, Kandu, a 5,000-pound female orca from Iceland, slammed into Corky, a 7,000-pound female orca from Canada, fracturing her own jaw and rupturing an artery in the process, causing blood to spurt through her blowhole like a geyser. While thousands of people looked on from the bleachers and the water in the tank turned crimson, Kandu bled to death. SeaWorld described the incident as a "normal, socially induced act of aggression." Scientists from the Humane Society disagreed: "It should be noted that two orcas from different oceans would never have been in such proximity naturally," they wrote in a report, "nor is there any record of an orca being killed in a similarly violent attack in the wild."

It's not only captive dolphins who get hurt by captive dolphins: people are set upon with some regularity. At SeaWorld Orlando's Dolphin Cove, a bottlenose chomped down on a seven-year-old boy's hand, clamping tenaciously even as two people tried to pry his jaws apart. One man had his sternum cracked during a dolphin encounter at another facility; in Japan, a woman had her back and ribs broken. Elsewhere, swim-with-dolphins clientele have had their teeth knocked out. In one recent episode in Curaçao, a bottlenose breached above a group of tourists, landing on them purposely. The more time you spend around pissed-off dolphins, the likelier your chances of trouble: in a survey conducted by the University of California, a full 52 percent of respondents who worked with marine mammals claimed to have been traumatically injured by them.

I came across a paper by a Russian scientist, G. A. Shurepova, that spelled out in graphic detail how these altercations can occur. He recounted one trainer's experience with a male bottlenose whom he had just separated from a group of ten other dolphins:

Trainer R first entered the pool to work with the animal. The dolphin swam up to him and took fish from his hands . . . The animal then abruptly veered to the side, it made two swift circles under the water, came up to the trainer, and plunged its snout in his side . . . Then followed a rain of gentler blows . . . The animal quickly turned about, hit the man painfully with the pectoral fin . . . It overtook the man and then turned to face him. The trainer halted, the animal rose vertically, put its head out of the water and dealt the trainer's body a blow with the caudal peduncle. The blow was very heavy and landed on the stomach, almost causing the man to lose consciousness.

Trainer R finally made it out of the water, but the dolphin wasn't finished. In subsequent training sessions, the dolphin "hit the man in the rib cage with its dorsal fin"; "thrashed vigorously in the water with its tail near his face"; "hit at the man's neck with its dorsal fin"; "struck him in the femur with its nose"; "hit him strongly in the stomach with its dorsal fin"; "lashed out at his arms with a sharp upward movement"; "gnashed its jaws menacingly, turned abruptly and gave a blow with its tail, but missed." The dolphin then "began to take bites at him."

An irate bottlenose can do some serious damage, but the dangers escalate with the animal's size: orcas, the largest species of dolphin, have their own, terrifying record: "Any person who has trained these animals has been thumped, bumped, bruised, bitten and otherwise abused over the course of time," one trainer told the *San Diego Union.* "It happens to everyone." A widely consulted training handbook listed "aggressive manifestations" of orcas as "butting, biting, grabbing, dunking and holding trainers on the bottom of pools and preventing their escape."

Not long before I visited the Dominican Republic, a harrowing incident had occurred in Florida, at SeaWorld Orlando, illustrating the ultimate hazards of dealing with a bored, frustrated, and manhandled whale. Tilikum, a 22-foot, 12,000-pound male orca, killed his trainer, Dawn Brancheau, during the "Dine with Shamu" show. As the audience watched in horror from a poolside café and through underwater viewing windows where families had gathered for a "Photo with Shamu," Tilikum seized Brancheau's arm in his mouth, pulling her from the deck into the water. While other SeaWorld employees did everything they could to save her—including dropping a net on the orca, and raising the bottom of the tank—Tilikum evaded them for forty-five minutes, shaking Brancheau, pinning her to the tank floor, breaking her neck and jaw, tearing off part of her scalp, severing her left arm, ending her life in a way that was hard not to see as completely intentional. This wasn't a freak occurrence, either. Only two months earlier, at a Canary Islands' facility called Loro Parque, trainer Alexis Martinez had been bitten, crushed, and drowned by an orca named Keto.

Though there has never been an instance of orcas fatally attacking humans in the wild, Brancheau was the third person who had been killed in Tilikum's tanks. During his earliest years as a captive, he was owned by Sealand of the Pacific, a marine park in Victoria, British Columbia. In 1991, one of his trainers there, twenty-year-old Keltie Byrne, had slipped while carrying a bucket of fish and fallen into his sea pen; Byrne died in much the same way Brancheau did. Although it is unclear whether Tilikum was solely responsible for her death—he shared the enclosure with two other orcas—he certainly participated. Tilikum was so reluctant to give up Byrne's body that it took rescuers almost two hours to retrieve it.

After Byrne's death, Sealand sold Tilikum to SeaWorld, which wanted a male orca for its captive breeding efforts, and he was moved to Orlando. Early one morning at his new home, security guards noticed Tilikum swimming around with something white flopped across his back; looking closer, they realized that it was a man's naked body. The man's name was Daniel Dukes. He was twenty-seven years old and a bit of a drifter, and he had made the ill-advised decision to sneak into Tilikum's tank for a dip one night after the park had closed. Divers were dispatched to pick up pieces of Dukes's body from the pool, including one of his testicles.

Even the most ardent fan of marine parks must see Tilikum's confinement as a tragedy, one with savage repercussions. He is an Icelandic orca, taken from his pod when he was two years old, and he has spent thirty years in captivity since then, at Sealand of the Pacific for seven years, and then at SeaWorld Orlando for the remainder. At both places, despite his moving-truck size, Tilikum was battered so badly by female orcas that he was often sequestered for his own safety.

Orca society is matriarchal and extremely tight-knit. In the wild, Tilikum would have spent his life with his mother. She would have taught him to speak the dialect unique to her pod, one that had been passed down through generations. He would have swum up to eighty miles a day in the rich, cold North Atlantic waters, and learned to navigate and hunt with an orca's masterful skill—they can take down gray whales with ease—and he would have sired calves out there too, mating with females from neighboring pods. Instead, he languishes in a solitary tank eating dead herring, and he is masturbated by SeaWorld staff wielding K-Y jelly, his semen used for the artificial insemination of other captives. Tilikum's real orca existence has been preempted, replaced by the Shamu sound track.

All attempts to domesticate the six-ton dolphin have failed. He is nobody's cartoon character, and yet all the ocean's magnificent possibilities are lost to him. So much of Tilikum's life has been spent among humans that he could never survive in the open sea. Even if by some miracle he were reunited with his pod, his teeth have been destroyed by years of chewing on his metal cage bars. They are drilled out and hollow now and would not be much use to him. He is the wildest of creatures, who will never get the chance to be wild. So what definition is left for Tilikum, caught between worlds? He has been reduced to the craziest possible hybrid: a serial killer used for entertainment purposes.

≋

I stayed for a while longer watching Serena and Niagara looping aimlessly, alone in their pool. The sun backed down and the afternoon light glinted apricot gold and the water took on a metallic sheen as the reflections sunk into shadow. Over by the lagoon I saw trainers striding briskly down the docks to feed the dolphins who hadn't performed in shows. The men all carried the same thrilling plastic cooler, and the dolphins leapt in their pens when they saw it. One dolphin rose out of the water and began to tail-walk, as if auditioning for fish. In a Jacuzzi-size enclosure, set off to the side, two Ocean World employees knelt next to a bottlenose who was floating quietly. A clear plastic tube ran from the dolphin's blowhole into a metal canister that was in turn plugged into an electrical socket, as though the animal were being vacuumed out. When one of the men turned and saw me looking at them, he signaled at me angrily to get lost.

I was startled by this frankly tourist-unfriendly gesture, and shifted so I was behind a sign and out of his line of sight. Serena swam over, idling in front of the railing where I stood. She lifted her head from the water and then rolled onto her side, as if trying to examine me from several angles. Unlike the chilled-out little spinners, these bottlenose dolphins seemed to be actively trying to breach the divide between their species and ours. I could see why the scientists who worked closely with them were moved by the experience, why John Lilly was so determined to find out exactly what was on their minds.

As I turned to leave, Alonso walked by. I nodded at him and started toward the exit, but then I remembered a question I had wanted to ask: "Hey," I called, getting his attention. "Where do these dolphins come from?" He stopped, and looked at me for a long moment. "Ah, that's a good question," he said. He was smiling, but his voice had an edge and his eyes were as hard as marbles. "We have one from Honduras, one from Cuba, and one from right over there." He waved vaguely toward the ocean. This did not add up to sixteen dolphins, obviously, but I didn't want to push it. I had asked a simple question, but, though I didn't know it at the time, a highly loaded one. Alonso's answer was an awkward attempt to evade it. What I also didn't know was that months later, when I learned the real stories of how dolphins are captured from the wild, they would pull me like an undertow, deep below the surface, into a world of nightmares.

CHAPTER 4

THE FRIENDLIES

On Ireland's Dingle Peninsula, a dramatic finger of County Kerry that juts into the Atlantic Ocean, the signs bear names like LIOS PÓIL and BALLY-FERRITER and FOILATRISNIG. There is Tralee, a musical-sounding place with sod-roofed hobbity houses and peaty bogs. Ballnavenooragh is known for its elaborate stone fort, abandoned in the thirteenth century. Riasc, a monastic ruin, is the site of beehive-shaped huts that are 1,400 years old. In Dingle, history lies in thick layers. The peninsula has Neolithic monuments and medieval castles, Bronze Age tools and Iron Age weapons, any kind of relic you can conjure, and there are mysterious rocks kicking around everywhere, carved with Celtic symbols and heaped into cairns and arranged in ritualistic ways like miniature Stonehenges. Every hamlet has its own ancient story, etched into the land. Every resident has a strong, proud sense of his past, commemorated daily with pints of Guinness.

Then there's the town of Dingle itself, which has Fungie the Dingle Dolphin.

I had heard rumors of Fungie, a male bottlenose who had forsaken the open sea to live inside the mouth of the Dingle harbor, a placid, shallowish inlet bordered by low verdant hills that are speckled with sheep. According to local legend, he had been swimming around in this area, not much bigger than a few city blocks, since October 1983. It did not seem like an auspicious place for a dolphin to settle. Though Dingle's bay is sheltered from snarlier North Atlantic conditions—churning seas, huffing winds—dolphins are well equipped for these things and seem to revel in the action: surfing down the faces of waves, leaping through the wakes of ships, playing in the maelstrom. By comparison, the Dingle harbor is a pond. After the dolphin, the next wildest animals in Dingle are cows.

Living inside the harbor would also expose Fungie to heavy boat traffic, the town's fleet of fishing trawlers motoring in and out, and the bay's tranquil appearance belied its extreme tides—some days its depth fluctuated as much as fifteen feet, a draining so drastic that if he wasn't careful a dolphin might easily end up beached in the mud. Dingle harbor couldn't be mistaken for a marine sanctuary either: it had been known in the past for its abundant reservoirs of trash. So what was a full-grown bottlenose with an entire ocean at his disposal doing in this fish tank? And where was his pod? Whenever Fungie was spotted—almost every day, apparently—he was always alone.

Despite the fact that Fungie has his own Facebook page and Twitter feed, I wasn't sure if he was quite so ubiquitous as I'd been led to believe. Somehow, the stories I'd read about him seemed more apocryphal than true. He struck me more as a town mascot than an actual animal, a kind of Gaelic Disney creation intended, most likely, to drum up tourism. For one thing, if the same adult bottlenose had resided in Dingle for the past thirty years, that would make him at least forty-two or forty-three years old, and though a fortysomething dolphin is well within the range of possibility, his uncommon living situation made longevity a long shot. Being part of a pod means protection, hunting success, society, sex, kin—the fundamentals of dolphin existence. A solitary dolphin is like a floating oxymoron. So how did this one survive?

The tales of Fungie the loner dolphin seemed improbable. But surprisingly, there are others like him. In fact, there are many accounts of dolphins who break away from their mates for reasons unknown—or are cast out or lost or orphaned or otherwise marooned—and end up seeking human companionship instead. These animals are typically bottlenoses, but there have also been instances of solitary orcas, spotted dolphins, common dolphins, dusky dolphins, Risso's dolphins, beluga whales, and even the rare tucuxi—a pink-bellied South American species—dwelling in a small area and community of people on a constant basis.

Scientists don't know why it happens, but tales of dolphins befriending humans reach far back into history. Aristotle wrote offhandedly about the dolphins' "passionate attachment to boys," as if everyone just knew this as a fact. In the year AD 77, Pliny the Elder, the early Roman naturalist and philosopher, recounted the story of a dolphin named Simo who formed a bond with a boy who fed him bits of bread: "At whatever hour of the day [the dolphin] might happen to be called by the boy, and although hidden and out of sight at the bottom of the water, he would instantly fly to the surface, and after feeding from his hand, would present his back for him to mount . . . sportively taking him up on his back, he would carry him over a wide expanse of sea . . . This happened for several years, until at last the boy happened to fall ill of some malady and died. The dolphin, however, still came to the spot as usual, with a sorrowful air and manifesting every sign of deep affliction, until at last, a thing of which no one felt the slightest doubt, he died purely of sorrow and regret."

Pliny goes on to chronicle more of these human-dolphin pairings, including the story of a dolphin who had frequently played with a Roman proconsul named Flavianus, only to refuse later to let the man ride on his back; for this snub the dolphin was put to death. In another instance, Alexander the Great had been so taken by the bond between a boy and a dolphin who followed him around, that he appointed the child High Priest of Neptune at Babylon.

When you consider how risky it is for dolphins to spend time in close proximity to people, it is all the more intriguing that so many human-dolphin stories have similar themes: dolphin seeks out man, dolphin wants to play with man, dolphin assists man, dolphin rescues man. If dolphins didn't already have such a well-established reputation for showing up like Superman in the third act, zooming in when people are in trouble, it would be impossible to put their behavior into context. But there are centuries and even millennia of tales of their generosity and kindness toward the awkward two-legged creatures they encounter who are so out of their element in the water. Dolphins have been known to respond like highly skilled lifeguards, saving people from all manner of aquatic peril. Occasionally they perform small kindnesses like retrieving lost diving gear or helping fishermen catch fish. It's hard to believe that dolphins actually care about us—"us" being the ones who ensnare them in nets and contaminate them with chemicals and make them do silly tricks and, in certain parts of the world, eat them— but at least some of them act like they do.

In the book *Beautiful Minds,* biologist Maddalena Bearzi recalls tailing a pod of bottlenoses on one grim, foggy morning along the coast of Los Angeles. The animals were hunting, ignoring her research boat as they searched for fish. Finally, they found a huge school of sardines and began herding them. If there's anything that commands a dolphin's attention it is a mother lode of fish, so Bearzi was surprised when one of the dolphins suddenly broke away from feeding and headed out to sea, swimming at top speed. The rest of the pod followed; so did Bearzi and her crew. The dolphins arrowed about three miles offshore and then they stopped, arranging themselves in a circle. In the center, the scientists were shocked to see a girl's body floating. She was a teenager and barely alive, her suicide attempt only moments away from succeeding. Around her neck, the girl had strung a plastic bag containing her identification and a farewell letter. Thanks to the dolphins, she was rescued. "I still think and dream about that cold day," Bearzi wrote, "and that tiny, pale girl lost in the ocean and found again for some inexplicable reason, by us, by the dolphins."

Tales like this are remarkably common: famously, when rescuers pulled five-year-old Cuban refugee Elián González out of the water three miles off Florida's coast, adrift and alone for forty-eight hours after his boat capsized and everyone else aboard had drowned, some of his first words were about how dolphins had surrounded him and kept him from slipping off his life ring in thirteen-foot seas. After a 9.1 earthquake shook the waters off Phuket, Thailand, on December 26, 2004, seven boats full of scuba divers were startled when a pod of dolphins began to leap theatrically, right in front of them, attention-getting behavior that none of the veteran captains or divemasters had witnessed before. The ocean was roiled from the earthquake and the captains had decided to return to port, but the dolphins seemed to be frantically beckoning them offshore. Out of curiosity, the dive boats followed them. At that point no one aboard the vessels could have known that massive tsunami waves were rolling beneath them, thundering toward the shoreline where they would cause epic death and destruction. The divers soon learned that by steering them out to sea, away from the breaking waves, the dolphins had saved their lives.

Surfers, in particular, seem to benefit from dolphin intervention; accounts of dolphins aiding their fellow wave riders are the most plentiful of all. When surfer Todd Endris was bitten three times by a great white shark near Monterey, California, dolphins drove off the marauder, formed a ring around Endris, and escorted him to the beach. Even before the attack the dolphins had been circling Endris with unusual focus. When the shark approached for its first pass they made a visible ruckus: thrashing their fins at the surface, slapping their tails on the water, and in general acting so aggressively that another surfer, Wes Williams, floating fifteen feet from Endris, wondered, "What did he do to piss off the dolphins?" When he got the full measure of what was happening, Williams paddled toward Endris to help. While Endris flailed in a pool of blood, the shark returned—twice. Williams watched one ninja dolphin vault out of the water and lash at the attacker with its tail, Bruce Lee style.

As a guest on *The Late Late Show* with Craig Ferguson, actor Dick Van Dyke recounted how dolphins had once pushed him back to shore after he fell asleep on his surfboard at Virginia Beach, accidentally drifting so far out into the ocean that he could no longer see land; Australian pro surfer Dave Rastovich, straddling his board waiting for a wave, was astonished to watch a dolphin hurtle itself at a shark that was torpedoing toward him, sending it fleeing. (Coincidentally, only two days earlier Rastovich had launched a nonprofit group, Surfers for Cetaceans, to protect dolphins and whales.)

What to make of these stories? One point worth noting is that dolphins often behave toward us in the same ways they do toward one another. It's standard dolphin operating procedure, for instance, to fend off sharks, or hold an injured mate at the surface so he can breathe, or steer the pod away from danger. In the dolphins' nomadic undersea world, solitude equals vulnerability, so a lone human in the water must seem to them direly in need of assistance. Their consideration of us isn't limited to emergency situations, either: at the Tangalooma Island Resort in Australia, where wild bottlenoses are regularly fed fish by people standing in the shallows, biologists have documented—on twenty-three occasions—the dolphins reciprocating, swimming up to offer freshly caught tuna, eels, and octopi as gifts.

In other words, dolphins do not always differentiate between us and them. At times they do not seem to care that we are not members of their species—they simply appoint us honorary dolphins. Maybe that was why Fungie had made his home among the residents of Dingle. To him, perhaps, they were just a slightly-peculiar-looking pod.

≋

I drove down to Dingle from Dublin, winding through green and peaceful country, through bustling little cities and quaint little towns, trying to get used to steering from the right side of the car and shifting gears with my left hand while driving on the left side of the road. It was automotive dyslexia, complicated by rain squalls, perplexing roundabouts, and narrow streets with mailboxes that all seemed to pop out at exactly the same height as my rearview mirrors. For six hours I managed not to sideswipe anything, and when I crested a final set of hills and dropped into the last valley before Dingle, I felt relief. Below me in the distance, I could see a platinum bay that changed color to bruised lavender as clouds lofted by overhead.

Dingle is a harbor town and everything in it points toward the water. Its main street fronts the bay itself, a soldierly row of crayon-colored pubs and restaurants and inns. From there the village lanes, postcard perfect in their charm, branch off like river tributaries, running slightly uphill. Fishing boats of various vintages are moored at the docks, two and three abreast, loaded with buoys and ropes and skeins of netting, jumbles of rusty parts that would be useful for something, if you could only remember what.

I parked my car near the town square and got out to take a look around. I knew I'd arrived at Dingle's hub because I was in a plaza paved with herringbone brick, spacious and inviting, and in the center, cast in gleaming bronze, stood a life-size statue of Fungie. Two children sat astride the dolphin eating croissants while their parents snapped pictures with their phones; other families hung back, waiting for their own photo opportunities. A small girl in a fuchsia tracksuit ran rings around the statue chanting, "Fungie! Fungie! Fungie!"

I stood there for a moment watching, then turned to a man holding a stroller next to me and asked if he had seen the dolphin in person. The man looked surprised. "Oh, yes. I've seen him many times," he said, pointing to the harbor. "He's always here." The fuchsia tracksuit girl came galloping over and stopped in front of us. Up close, I could see that she had a blue dolphin tattoo on her cheek—temporary, I hoped. "Can you tell her what the dolphin did today, Clare?" the man asked brightly, with an aside to me in a lower voice: "Scared the living daylights out of her." The girl nodded vigorously: "He nearly touched Daddy's head!" I found this hard to visualize, but before I could ask for clarification the man had turned away and his daughter clambered aboard the statue, shoving aside a boy who was reclining on Fungie's tail fin.

Behind the dolphin there was a stone building that looked like a harbormaster's office; its windows were plastered with Fungie posters and advertisements and press clippings. I wandered over to read them. An eight-foot cardboard cutout of the dolphin reared up behind the glass: FUNGIE IS FINTASTIC! COME SEE HIM TODAY! INFO@DINGLE DOLPHIN.COM. On an illustrated map of the surrounding waters, Fungie had been painted erupting from the center of the bay, labeled in Irish as "Doilphín." A newspaper photo showed him lolling on his back next to a red skiff, while a woman leaned over the side to tickle his belly. DOLPHIN WHO JUST WANTS TO HAVE FUN, read the headline. So why had the "the big-hearted bottlenose" stayed in Dingle for so long? The writer had asked locals for their opinions. "Maybe he's just very sociable and likes the Irish way of life," one resident mused.

Another article announced that "fun-loving Fungie the dolphin has somersaulted into the record books . . . as The Most Loyal Animal on the Planet!" To win this title, I read, Fungie had outdone a Risso's dolphin named Pelorus Jack who spent twenty-four years, from 1888 to 1912, escorting ships through New Zealand's Cook Strait, a tricky slice of sea between the North and South Islands. These waters contain every possible treachery: rough waters, submerged rocks, whipping winds, and a fierce current known as Te-Aumiti—Swirling Vortex—by the Maoris. Before the dolphin stepped in, the Cook Strait had hosted a number of New Zealand's worst maritime disasters.

Pelorus Jack's job, as he performed it, was to guide boats to a safe crossing. Usually he would just materialize at the bow; if he didn't immediately show up, captains would often stall their vessels and wait for him. For his navigational expertise and the joyful way he expressed himself—jumping across bow waves and rubbing up against ships' hulls—the dolphin was beloved. "He swam alongside in a kind of snuggling-up attitude," one seaman recalled.

By all accounts Pelorus Jack was a handsome animal, about fourteen feet long and colored a mottled silver, darker at the tips of his fins. As he got older, he turned white and his countless scrapes and scratches and scuffs stood out in relief, as though he had been the target of intensive graffiti. Like all Risso's dolphins, he had a large round head and a tiny snip of a beak, giving him a wry, brainy appearance. During his tenure, Pelorus Jack's reputation spread far. Tourists flocked to see him. Songs were written about him. Rudyard Kipling and Mark Twain both watched him in action. Sometimes he appeared in gossip columns.

When a passenger aboard a local ferry, the *Penguin,* shot at Pelorus Jack with a rifle, grazing him and causing him to disappear for weeks, the New Zealand government passed a national law specifically to protect him. After his gunshot wound healed, the dolphin returned to his post. But, numerous witnesses claim, he never guided the *Penguin* again. If he happened to see that particular ferry he would dive immediately and vanish. In a twist of history that might be described as karmic justice, three years after the incident the *Penguin* hit a rock and sank.

~~~

The next morning I bought a ticket for the *Lady Avalon,* a sturdy blue and white trawler that departed at nine a.m. for a Fungie tour of Dingle Bay. It was an overcast day, windless and soggy with moisture. The air tasted like brine. I zipped up my raincoat and stood, as directed, on the boat ramp near a sign that said WAIT BEHIND LINE FOR DOLPHN FERRY. I had wondered if the weather would deter dolphin watchers, but nobody seemed to care or even notice the drizzle. Young families swaddled in Gore-Tex, and outdoorsy couples cradling cups of coffee, lined up beside me.

When you consider the dolphin-themed economy of Dingle, with its stores selling dolphin postcards and dolphin T-shirts and dolphin earrings, its art galleries full of dolphin paintings and dolphin drawings, its pizzerias serving Pizza Fungie—the entire enterprise of greater Dingle—it is clear this is the town that Fungie built. According to Dingle's tourist bureau, 75 percent of all visitors come for one reason: to see its resident dolphin. During the summer, that influx adds up to about five thousand people per day. Including the *Lady Avalon,* there are nine dolphin ferries, and every one of them makes multiple trips each day, seven days a week, an hour on the water with a money-back guarantee if Fungie does not appear. A single ticket costs sixteen euros and each boat holds at least thirty people. I did some quick math.

In a 1991 documentary titled *The Dolphin's Gift,* filmmakers interviewed the town's elders about their impressions of Fungie, as if trying to gain insider information about what on earth the dolphin was doing there. One scene takes place in a pub. Watching it, I had the impression of peering into Dingle's civic soul. An old man with a florid drinking complexion, alarming growths on his nose, and wild tufts of pale hair leans into the camera clutching his pint. "The dolphin is a gold mine," he says, in a thick Irish brogue, as fiddle music wails in the background. "Do you know what this is? A gold mine." He crosses his arms, and laughs throatily. "Ohhhh, yes. I'm very much in love with the dolphin."

A deckhand ushered us aboard and within minutes we were motoring slowly away from the dock. Dueling cormorants flew loops overhead. In the near distance I could make out the bay's pinched entrance, flanked by sloping headlands. The water was a swampy olive color, flat as a bathtub, but when the light glanced off it a certain way it turned to an upbeat slate blue. A dinghy, a white Zodiac, and a trio of sailboats circled in the center of the bay. The sailboats were kid-size, barely big enough to hold two people; one of them had tomato-red sails that stood out like flags against the emerald fields and hills. Every few minutes the skipper of this sailboat leaned over the side to whap the water with something that looked like a spatula. I bent myself nearly double over the railing and stared down into the bay, but I saw nothing. If it were not for the knowledge that a large animal was hiding somewhere in here and that he might pop up at any minute, this harbor circuit would easily qualify as the world's most boring boat ride.

"Anyone see him anywhere?" The captain, a man named Jimmy Flannery Sr., stuck his head out of the wheelhouse. No one had, but not for lack of looking. People were stacked on top of one another with their cameras trained on the water. Every slight ripple or movement caused the flotilla of tiny sailboats to go tearing off in one direction or another. "Where is the dolphin?" a girl asked loudly. "Where is he, Mum?" Her mother lifted her up so she would have a better view. "Emily, you have to watch out," she said, "because the dolphin *is* out there somewhere."

Suddenly, from the stern, a lady in a yellow slicker yelled: "There he is! Oh my God! I saw him!" With a whooshing outbreath the dolphin had surfaced, and he was close enough that I could see his distinctive, gnarled face. Fungie looked pugilistic, and disconcertingly huge, with white markings around his chin like an old man's whiskers. He bore noticeable scars: his beak was roughed-up at the tip and his tail was missing a divot. On his throat he had the dolphin equivalent of deep wrinkles. Still, this was a big, tough bottlenose. I had read that Fungie was twelve feet long and weighed seven hundred pounds, but those numbers are low. My first thought was that the Most Loyal Animal on the Planet could knock someone's lights out if he wanted to.

After following alongside us Fungie dove, rocketing into the air seconds later next to the sailboats. He seemed especially fond of the boat with the red sails. Fungie spy-hopped and lunged out of the water and splashed the sailors with enthusiasm. "It's sport for him," a bearded man in a checked cap said to me, nodding in Fungie's direction. "Rogue dolphin, he is. They leave their flock."

Watching the dolphin, I felt a palpable glee, a doglike joy, emanating from him. No wonder the town had claimed him as their own— Fungie is as recognizable as any person. His face is as unique as yours or mine; in photographs it is obvious that he has aged over time. To see him is to know with certainty that he is an individual with his own quirks and traits and habits, his own way of presenting himself in the world. There is absolutely no way that the Dingle Chamber of Commerce could surreptitiously swap in another dolphin as a substitute for Fungie, should he ever fail to report for boat-playing duty.

Once he'd arrived at our party Fungie was a skilled entertainer. He made perfect aerial arcs and walked on his tail and at one point he swam along on his back, clapping his pectoral fins. Many of his moves were surprisingly showy, less like natural behaviors than tricks he might have been taught in some lost chapter of his pre-Dingle life. Observing him, I found myself wondering if Fungie's past might have included a stint in captivity; if, back in the day, he had lived in a sea pen and somehow escaped. It had been known to happen, especially during storms. In Hurricane Katrina, for instance, eight dolphins from the Marine Life Oceanarium in Gulfport, Mississippi, were swept from their pool by a thirty-foot storm surge, and landed in the Gulf of Mexico. Those dolphins ended up back in custody, but on occasion captives do get away. Unfortunately, they don't always know where to go or what to do with their sudden freedom, and so they seek out what they've become accustomed to: people. Could Fungie be a refugee?

We can only guess. Fungie's early life history is irretrievable, erased by time and myth. Like every solitary dolphin, he comes with a built-in mystery: How did this happen? When, and why? Getting entangled in nets or fishing lines for a time, losing contact with his pod in rough seas, becoming orphaned or sick or injured for whatever reason—any of these situations might strand an individual dolphin, leaving him to fend for himself. Or, for all we know, Fungie may have swum into this harbor, liked it, and simply decided to stay.

It was the Dingle fishermen who had noticed the dolphin first, and given him his name. Fungie trailed behind their boats as they returned to port, hoping, no doubt, for a handout of fish, but probably also yearning for company. At sunset he could often be seen jumping in the center of the bay, framed in silhouette like a dolphin on a movie poster. Sometimes, as if to show his gratitude or establish himself as a thoughtful neighbor, Fungie would catch pollack, salmon, and trout, and toss them into boats.

Three years into his Dingle tenancy the dolphin got some swimming companions. Sheila Stokes and Brian Holmes, a couple from nearby Cork, showed up in thick wetsuits, slipped into the water, and began to snorkel with him. For weeks Fungie orbited them at a distance, but Stokes and Holmes were persistent, spending hours in the frigid bay. They were respectful, too, letting the dolphin initiate all contact. Their patience paid off: Fungie began to brush against Stokes's outstretched hand. "You could sense his excitement, as well as my own," Stokes said, "because he went off and did a few leaps and flips in the bay before coming back for more touching. And from then on, he let us touch him a lot of the time." While Stokes rubbed Fungie's fins and belly, and ran her hands over his beak and his head, Holmes shot video of the dolphin looking as moony as a high school kid with a crush.

These images trickled out, followed by some local press. Soon a steady stream of people flowed into Dingle, eager to have their own transcendent dolphin encounters. In general Fungie handled it well, allowing quite a bit of interaction. He had his favorites, certain swimmers or kayaks or boats that he preferred, though whenever Stokes showed up it was as if no one else existed. The bay filled with sailors and snorkelers and diving groups, children bobbing in lifejackets, teenagers gunning Jet Skis, people rattling anchor chains and towing boogie boards to get the dolphin's attention. Typically Fungie responded to chaos or belligerent behavior by sensibly swimming off, but one time he rammed a German tourist in the groin, an injury that sent the man to the hospital.

These days, with the entire membership of the Dingle Boatmen's Association operating dolphin-watching tours, plus a fleet of recreational boats plying the harbor, the swimmers had for the most part given up. In any case, Fungie seemed to appreciate speed. When Flannery opened the *Lady Avalon*'s throttle, the dolphin careened away from the red sailboat and began to porpoise alongside us, sailing as high as the railings. "Wahhhhhh!" yelled a boy, as Fungie burst out of the water only inches away. By now our hour was almost up; it was as though Flannery and Fungie had planned this series of leaps as a grand finale. It could not have been executed any better if it had been part of a Vegas show.

Back at the docks, I asked Flannery which among Fungie's repertoire of tricks was the most impressive, whether the dolphin had ever amazed him with some improvised move. The captain scratched his head beneath his cap and nodded: "He does a backflip. Comes clear out of the water."

I decided to share my theory, not realizing that by doing so I was committing Dingle heresy. "It seems like someone must have trained him," I said. "Do you know if they did?" Flannery, who had been smiling pleasantly enough before I said this, turned and stared at me hard. A shadow passed fast over his face, darkening it like a thunder-cloud. "Not at all," he said curtly, turning away dismissively. "He is a totally wild animal."

≋

Dolly in France and Paquito in the Basque country; Egypt's Olin, who befriended a tribe of Bedouins in the Gulf of Aqaba; Charlie-Bubbles from Newfoundland; Springer from Seattle and Scar from New Zealand; Chas, who loved a particular buoy in the Thames—these and so many other solitary dolphins have made themselves known to us. That is usually where the problems begin.

The inevitable unruly relationship between a solitary dolphin and the people who want to see him vexes biologists, who fear—correctly—that these encounters will end badly for the dolphin. In this case Fungie is an exception, having far exceeded the life expectancy for a wild bottlenose who interacts daily with humans. Sadly, most friendly solitary dolphins don't last for even a fraction as long. Their biggest threat, by far, is propellers, which seem as alluring to curious dolphins as they are deadly: scientists have heard dolphins playfully mimicking the sounds of motorboat engines underwater, the way children do with their favorite toy trucks.

Wilma and Echo, orphan belugas from Nova Scotia, both died from propeller strikes, but not before charming thousands of people, gliding up to sightseeing boats to let passengers stroke their skin. Jet, a bottlenose from the Isle of Wight, had his tail lopped off by a propeller and bled to death. Freddie, a bottlenose from Northumberland, U.K., whose companion had swallowed a plastic bag and washed up dead on the beach, liked to swim upside down beneath motorized dinghies; he also enjoyed the flume of a sewage outtake pipe. Both were dangerous attractions. The chemical-treated waste infected Freddie's skin, turning it a grizzled white; but once again, it was a propeller that got him.

Luna, an endearing lone orca calf who lived near a marina in Nootka Sound, British Columbia, was the subject of a movie, *The Whale*, narrated by Scarlett Johansson. He survived for five years before being hit by a tugboat.

JoJo, a bottlenose celebrity from the Turks and Caicos Islands—the country's government officially declared him a "national treasure"—appears to hold the record for propeller mishaps. In a single seven-year span, from 1992 to 1999, he sustained thirty-seven boat-related injuries, eight of them life-threatening. (Against the odds, JoJo made it through all that and continues to visit his usual headquarters, the waters near Club Med in Providenciales.)

But propellers are only one hazard among many. To read through a list of friendly wild dolphins who have met violent and untimely ends is to read a list of appalling human behavior. Over in Israel, Dobbie, a bottlenose who liked to play with the air bubbles from scuba divers, washed up full of bullet holes. In Australia, Zero Three, a young male bottlenose, was poisoned by toxic chemicals that were dumped in the river where he swam. The Costa Rican, another bottlenose, fell in love with a local dog whom he would meet every day, pushed children around in a canoe, and let people ride on his back. When he became entangled in a fisherman's net, he waited calmly for help to arrive. Instead, the fisherman gaffed him and dragged his body onto the beach. A French bottlenose named Jean Floc'h paid for his fascination with rowboats: he was beaten savagely with wooden oars. Dolphins who gravitated toward us have also been stabbed with knives and screwdrivers and even ballpoint pens, garroted by wire and fishing line, pierced by spearguns, targeted with explosives, and purposely run over by Jet Skis.

Surrounded by people who want to swim with them, touch them, and grab their fins, dolphins can become aggressive themselves. Lone dolphins, removed from everything familiar and confused by their new acquaintances, have been known to pin snorkelers to the seafloor, break arms, ribs, and noses with their beaks, make amorous advances, and club swimmers with their tails. Far from rescuing people, if a dolphin is sufficiently riled-up he might prevent them from exiting the water, or push them farther out to sea.

One fatal human-dolphin clash took place in São Sebastião, Brazil: a solitary bottlenose named Tião had been plagued by crowds of people dropping popsicle sticks into his blowhole and pouring beer into his mouth. Eventually the dolphin had endured enough. He sent twenty-eight of his tormentors to the hospital, but still the harassment continued. When two drunken men tried to wrestle Tião into the shallows so they could have their picture taken with him, he walloped them with his fins and tail. Both men were injured; one later died from a ruptured spleen. KILLER DOLPHIN! screamed the headline in the local paper. Not long after that, Tião disappeared and was never seen again, presumed killed in retaliation.

Lakeshore Estates, a gated waterfront community in Slidell, Louisiana, dealt more humanely with its hostile resident bottlenose, known impersonally as The Dolphin. During Hurricane Katrina the young male bottlenose had gotten separated from his pod, and ended up alone in a brackish canal in the middle of the suburb. In the seven years since his arrival, The Dolphin had done quite a bit of damage, and his behavior was becoming increasingly ornery. In a flurry of activity he'd bitten several people—including one girl he'd attempted to drag away from shore by the ankle—chased swimmers out of the water, snapped his jaws at kayakers, and body-slammed dinghies. "Slidell Memorial Hospital's press secretary could not be reached to confirm how many dolphin attacks the hospital has recently seen," reported the *St. Tammany News* from New Orleans.

Concerned about The Dolphin's surliness, residents held a community meeting with biologists from the Louisiana Department of Wildlife and Fisheries, and the National Oceanic and Atmospheric Administration (NOAA). The meeting's official title was "The Slidell

Dolphin Challenge: *What Can We Do?*" About sixty Slidell locals attended, along with a pair of sheriffs. "This thing has been a problem for years," a thin man with a bushy white mustache complained. "Why can't we remove it? You know if you put it in an aquarium, the problem is solved."

"Or maybe they should find him a girlfriend," suggested a woman in a white pantsuit and red lipstick. (It was true that The Dolphin often swam around with an erection, which he rubbed against boats.)

"The problem is the people," a burly Cajun wearing a Coast Guard baseball cap shot back, looking at Mustache Man sternly. The biologists agreed: the best thing the community could do for The Dolphin was to steer clear of him. No more racing him on Jet Skis. No more feeding him bits of hot dog. No more attempts to pet him, no more following him around to take smartphone videos of his penis. The less human interaction he had, the crowd was told, the better his chances of survival. If he were ignored, The Dolphin's bad behavior—at least the infractions directed toward his two-legged neighbors—would most likely stop.

Aside from some mild grumbling, the Slidell community seemed to understand that, and even to empathize with The Dolphin's circumstances. They worried aloud about The Dolphin getting shot or otherwise harmed as a result of his recalcitrance. "You know, he's just like us," said another man, whose home and business had been dashed by the hurricane. "He lost everything, but he's put it behind him and is fine. He's a survivor. People just have to leave him alone now."

≈

The one thing we know for sure about lone friendly dolphins is that we are likely to meet more of them. A 2008 global census of wild dolphins who have sought human company charts a dramatic rise since 1980. In their book *Dolphin Mysteries,* researchers Kathleen Dudzinski and Toni Frohoff note that "with increasing frequency we are seeing greater numbers of solitary, sociable toothed whales." Across the world, it seems, their society and ours are colliding.

When you think about it, this culture clash is inevitable. A dolphin doesn't end up with a Twitter account unless something has gone very wrong before that, and the oceans these days are a mad mess of trouble. Even if dolphins manage to evade our web of fishing nets and longlines, they still contend with relentless pollution, oil spills, habitat destruction, food depletion, a barrage of brain-jangling noise—the list goes on. Of course we'll find them among us: they have nowhere else to go.

In so many ways, I came to realize, Dingle is a best-case scenario for a podless dolphin. There is no way to watch Fungie and doubt that he is having fun. He hunts for his own food. He is savvy enough to avoid propellers and discerning enough to dodge assholes. He has bonded with people but he's not completely isolated from his own species: at times, other dolphins venture into the bay. Lately, Fungie has been seen gallivanting with two female bottlenoses, the three of them leaping in tandem and appearing to embrace one another. Fungie, the local newspaper bragged, "is something of a ladies man." In all situations the town protects his interests: what's good for Fungie is good for them. And if The Most Loyal Animal on the Planet ever decides that he has been loyal to Dingle for long enough, he is free to leave as he pleases.

Obviously, the town is praying that never happens. In 2013, to mark the dolphin's thirtieth anniversary as a citizen, Dingle threw a three-day party, the Fungie Festival. Reading the schedule, I almost fell over in delight. Events included art and photo exhibitions (images of Fungie), poetry readings (works inspired by Fungie), concerts (music written for Fungie), historical lectures (Fungie: The Early Years), scientific talks (Fungie and Other Solitary Dolphins Around the World), children's book readings (a series starring Fungie), conversation circles at village pubs (people talking about how Fungie has affected their lives)—plus morning swims, evening swims, and a boat convoy to bless the dolphin "out of gratitude for all he offers so freely to folks and to celebrate his presence at the mouth of the harbor."

As I drove away from Dingle, the bay shining behind me, I gave my own silent thanks to the people who had cared enough to protect a lone dolphin, the town with a Fungie-shaped space in its heart. The dolphin has repaid them in kind, and then some. It's an uncommon relationship and beautiful to see. I planned to keep the Fungie Festival, all the brightness of that gathering, in the front row of my mind: I knew I would need it as a talisman. The next place I planned to visit was also a picturesque fishing town. It, too, had dolphins. There, as well, dolphins played a major role in local affairs.

But instead of being a haven for the animals, this town had chosen differently.

CHAPTER 5

WELCOME TO TAIJI

"Okay now. We've just been told that we need to go directly to the police station for processing." Mark Palmer stood at the front of the bus to address us. He was sweating and smiling. I liked Palmer's voice, a reassuring baritone uplifted by a cheerful note of subversion. *We may be headed into trouble*, his tone implied, *but we will have an excellent time*. Actually, I liked everything about Palmer—and his colleague Mark Berman and, in fact, everyone aboard this tour bus with its blacked-out windows and cushy seats, winding its way through the cedar-cloaked mountains and around the coastal S-curves of the Wakayama prefecture in southeastern Japan, rolling toward our destination: the notorious dolphin-hunting town of Taiji. What I didn't like that much was the idea of heading straight to the authorities when we got there.

If you've seen the Oscar-winning documentary *The Cove*, you've had a glimpse of what happens in Taiji. This pretty seaside town of 3,500 people is up to some very ugly business: catching, killing, and selling dolphins. Before they hunted dolphins, the Taiji fishermen hunted whales. Their whaling history dates back to 1675, but their dolphin hunt, which began in 1969, is relatively recent. It is also mercenary, environmentally disastrous, and exceptionally brutal, and it has drawn worldwide howls of protest. In response, the town is hostile to outsiders. To the Taiji dolphin hunters—and the local people who support them, the politicians who protect them, the dolphin traffickers and yakuza gangsters who profit from them, the marine parks that buy dolphins from them—anyone coming to protest the hunt falls somewhere on a sliding scale between irritant and terrorist. This is demonstrated clearly in *The Cove*: in Taiji, the film's main character, American activist Ric O'Barry, is regularly followed, screamed at, ejected from public places, and threatened. "They'd kill me if they could," O'Barry says.

O'Barry, now seventy-five, is a famous crusader against dolphin injustice. He has a unique résumé: he is Flipper's former trainer. Throughout the sixties, O'Barry taught the five female bottlenoses who played Flipper to perform the dazzling feats that made the TV show such a hit. Before that, he collected wild dolphins for the Miami Seaquarium, scooping up more than a hundred animals around Florida and the Bahamas. O'Barry caught dolphins for captivity and looked after dolphins in captivity and trained dolphins in captivity, and he lived large while he was doing it, cruising around Coconut Grove in a Porsche, entertaining models and rock stars who dropped by to hang out with Flipper, pulling down a hefty salary. There was only one problem. Working with the dolphins, O'Barry began to have the same weird feeling that John Lilly had described: "About halfway through the TV series I really started having second thoughts about captivity. But I didn't actually do anything. Things were going too well to ruin the party." The turning point came in 1970 when his favorite dolphin, Kathy, died in his arms in a seemingly intentional way—looking up at him and simply refusing to take another breath. "She was really depressed," O'Barry recalled. "I could feel it. I could see it." The dolphin, he believed, had committed suicide. The next day, O'Barry switched careers. For the past forty-three years he has dedicated his life to dolphin welfare, roaming the globe doing everything he can to help them.

It was O'Barry who had invited me to Taiji. Each year on September 1, opening day of Japan's six-month dolphin-hunting season, O'Barry and his group, the Dolphin Project, hold a vigil in the cove, and then every day after that they track the hunters' movements, documenting everything that occurs. "I will keep returning to Taiji until they stop or I drop," O'Barry told me. I had instantly agreed to come, though I knew the trip would be disturbing. I wanted to experience the place for myself. Emotions run hot around dolphins wherever they are, but in Taiji everything is taken to an extreme.

I flew to Osaka on August 30, but at the last moment O'Barry was summoned to appear in court on a dolphin-related matter and had to delay his arrival. It was decided the rest of us would push on to Taiji as planned; O'Barry would join us two days later. Along with me there were thirty others on the bus, an international mix of activists mostly in their twenties and thirties, three translators, plus Palmer and Berman. Both men were associate directors at Earth Island Institute, the environmental organization in Berkeley, California, that worked with O'Barry; both had plenty of experience dealing with wrathful fishermen and fraught politics and smiting rage aimed in their direction, all of which we were likely to encounter.

As we approached the town's outskirts, Berman, sitting next to me, leaned back in his seat, shook his head glumly, and crossed his arms, as if battening down his personal hatches. "We're about ten minutes away from the cove," he said. "Once you get there you just feel really sick." I nodded, though it was hard to imagine Berman buckling that quickly. During the six-hour drive from Osaka I had heard stories of his exploits and decided that if I ever needed to do something highly unpopular and unpleasant, I'd call Berman for advice. His nickname is the Berminator.

With his wire-rimmed glasses, gray hair, and small build, Berman looks more like a high-school math teacher than the seasoned eco-streetfighter that he is. At Earth Island he heads the Dolphin Safe tuna monitoring program, a watchdog initiative started in 1990 to prevent fishing fleets from snaring dolphins in their catch. In this role Berman must enforce strict regulations on even the most piratical operators, none of whom give a damn about dolphins but all of whom cannot afford to have their tuna boycotted by dolphin-loving consumers. Wielding this economic threat does not always make him an excess of friends. Berman has been lunged at by Solomon Islands tribesmen, assaulted by a fishing crew in the Philippines, and told by a Thai businessman, "You know, people like you can disappear very easily." None of it seemed to faze him. "I grew up as a Jew with curly hair in the deep South in the sixties," he said, explaining his dauntlessness. "That was pretty rough."

Through the bus windows I could see a hillside tumble of closely packed houses with brown and beige and rust-colored roofs, somber earth tones rather than happy seaside pastels. We drove by a tsunami warning sign, a yellow triangle filled with ominous black waves, and then we crossed over a bridge that was topped by a dolphin statuette. The dolphin was the size of a garden gnome, and its mouth gaped open as though it was yelling for help. We passed beneath full-scale models of a humpback whale and her calf, impaled on steel pillars so they appeared to be flying through the air.

In Taiji central we were greeted by the *Kyo Maru I,* a dry-docked whaling ship. The 800-ton vessel was painted battleship gray and dull red, with the word RESEARCH splayed boldly across its hull. Not long ago it had chased whales in Antarctica; now it was sidelined, raised up on concrete blocks like a monument. The world had acted on behalf of the great whales, banning commercial whaling in 1986, but small whales were not specifically included in the agreement. Taiji and several other Japanese towns kill up to 20,000 dolphins each year, taking advantage of this omission.

In general, there is a fervent difference of opinion between Japan and other nations about whether *any* cetaceans deserve protection. Japanese fishery officials have openly described killing dolphins and whales as something akin to a public service; one spokesman referred to them as "the cockroaches of the sea." In their upside-down equation, fewer cetaceans add up to more fish. In truth, the ocean is more complicated than that, and removing predators from its waters disrupts an intricate balance. Scientists now know that cetaceans not only help maintain healthy fish populations, they play an essential role in creating them.

Whale and dolphin meat—illegal to eat in most countries under the Convention on International Trade in Endangered Species (CITES)—is an entrenched part of Japan's national cuisine. In Taiji, the *Kyo Maru I* sits as a proud emblem of the country's whaling expertise; to me it was a haunting reminder that we almost harpooned the animals out of existence. I looked past the ship to the Pacific, tranquil and glassy in the early evening light, and then we rounded one last corner and the cove lay in front of us.

It was smaller than I'd realized, maybe two hundred yards long and sixty across, U-shaped and studded with rocks, lined on both sides by steep, thickly forested bluffs. From the road you could look down onto the cove's pebbly beach and into its shallow green waters. The place was eerily still, without even a breath of wind. A white buoy line stretched from one side to the other, though nobody was swimming. A picnicking pavilion stood deserted. The streets were empty; the seaside walkways were blocked by heavy chain-link and barbed wire fencing with signs that said: NO TRESPASSING. DANGER. FOR A FALLING ROCK and THIS IS A RESTRICTED AREA. At the cove there were no children playing with toy boats, no Frisbee-chasing dogs, no summer ice cream vendors, no couples enjoying the sunset. If you didn't count the forty policemen standing across the street, in fact, there was no one here at all.

Our bus driver turned into a parking lot, and stopped beside some lonely-looking palm trees. This was the cove's new police station, built the previous year. Altercations between dolphin activists and dolphin hunters could get heated, and the number of people who showed up here each September 1 kept rising. Lately, ultra-right-wing nationalists and yakuza had come around spoiling for a fight; neither was a threat to be ignored. In past months, protesters from the group Sea Shepherd had been tailed and harassed; there had even been an attempted abduction. Responding to the ratcheting tensions, the federal police had become a constant presence in Taiji. O'Barry's groups were always peaceful and law-abiding, but even so, the police wanted to interview us individually before we could check in to our hotel.

Palmer stood up again. "We'll go in two at a time," he said, as the bus doors opened.

"Be pleasant. That means you, Berman." He chuckled dryly. "And remember, if you don't understand what they're saying, just smile and nod."

≈≈≈

"I ask you some questions so please." The policeman was businesslike. He sat behind a desk wearing a neat navy uniform and a poker face. His English was halting but workable. Opening my passport, he examined it for what seemed like a very long time, making careful notes as he flipped through its pages. After a while he leaned over to another policeman sitting next to him and pointed to one of my visa stamps. The two men began to speak in rapid-fire Japanese. I smiled and nodded.

The first policeman turned back to me. "What are you doing here?" he asked.

"Uh . . ." Was there a right answer? I decided to go for diplomacy: "I'm here to learn."

"Ohhhhhhh."

"I observe."

"Mmmmmm. Demonstration is here tomorrow. Do you take part in this?"

"Yes."

"Oh oh oh oh oh. Tell me about one thing. Have you ever heard of the conflict right wing nationalists?"

"I heard something."

"They are waiting in the next town," he said, shaking his head and looking grave. "They can be dangerous. They are gonna come here tomorrow, many. Say bad things. So please be careful."

"I will. I will be careful."

"Please ignore."

"I will, definitely."

"And don't push anyone."

"No, no. I don't push."

He handed back my passport and motioned for the next person to come in. Nodding *arigato,* I went outside to join the others. Veronica, a graceful Bolivian who had traveled here with her daughter, was staring at the cove, crying. Two Australian girls, Yaz and Britt, both from Perth and just shy of twenty, stood next to her. "I can't take it all in," Britt said, looking stricken. She was no lightweight in the compassion department: at fourteen, she had saved up her money and flown to Nepal to work at Mother Teresa's mission.

A young Californian named Arielle walked by, dripping saltwater with every step. She had waded into the cove and dunked herself and now she was ready to answer their questions. Her soaking wet clothes clung to her small frame; mascara ran down her cheeks. Arielle had a long shag haircut and a round, knowing face. She was a singer and planned to perform tomorrow during the demonstration. On the bus I had watched her lovingly tuck her guitar into its case; the front of the instrument was coated with fluffy white feathers. As Arielle opened the door to the police station, Berman exited. He did a double take as she passed. "Oh man," he said. "That takes some guts. To get in *that* water."

We all knew what he meant. It's the rare seaside cove that doesn't invite dipping, but splashing around in here would be like picnicking in an abattoir. Even worse than the knowledge that tens of thousands of dolphins have died in this spot are the specifics of how they were killed. When the dolphins are driven into the cove, they're disoriented and terrified. Sometimes they are left in there for days without food or hydration (like us, dolphins need fresh water; they extract it from the fish they eat). When a slaughter begins, the pod can hear one another's cries. They know exactly what is going on. Clandestine footage of past hunts reveals a breathtaking amount of casual cruelty from the fishermen—dragging the dolphins out of the ocean by their tails, piling wounded dolphins on top of each other, stepping on the dolphins—laughing and smoking while they're doing it.

Recently, the Taiji fishermen introduced a new technique for killing dolphins—which they claim is far more efficient and humane—and it involves severing the animals' spines while they are alive, spiking metal rods into their blowholes, and then plugging them with wooden dowels. Death is far from instant: first the dolphins endure paralysis, then they die gradually of shock, hemorrhage, drowning, or asphyxiation. This practice is so merciless that a group of scientists and veterinarians in Britain and the U.S. released a paper analyzing the resulting trauma; they concluded that "this killing method . . . would not be tolerated or permitted in any regulated slaughterhouse process in the developed world."

≈

Interrogation over, we drove to the Kayu hotel. It was a blockish but well-situated affair, only a stone's throw from the cove. We were spending a single night at this hotel; for the rest of the week we would stay in Kii-Katsuura, the neighboring town, five minutes away. As our luggage was unloaded, I asked Berman why we had to move. He grimaced. "Staying in Taiji is really not fun," he said. "They don't want us here. They'll take our money, but they really don't like us." "Oh, they don't like us *at all*," a slender blond woman standing next to him added. "Kii-Katsuura is bigger and the hotel there has better security."

The woman's name was Carrie Burns, and she spoke from experience. Carrie and her husband, Tim, had been regulars in Taiji for the past three years, flying from their home in St. Petersburg, Florida, and often staying for weeks at a time. Like everyone else in the group, they were propelled here after seeing *The Cove*. "We watched it on Netflix," Carrie said, "and that same night we were online, researching the price of tickets to Japan." Once alerted, the Burnses did not commit themselves halfway. Carrie had learned the hunt's every machination, down to the quotas for each dolphin species, and could discuss any aspect of it with encyclopedic recall. Tim ran O'Barry's cove-monitor program, making sure that each morning at five o'clock when the dolphin-hunting boats left the harbor, someone was perched at a hillside lookout with a set of binoculars and a camera with a long lens. Whenever dolphins were driven into the cove, the monitors took pictures and video, counted the captured animals, surveilled everything that was happening, and then broadcast it all through social media.

As images streamed out of the cove's water stained red with blood, of fishermen hacking at dolphins with spears and knives, and dolphin trainers wading in to select the youngest and prettiest of the captives—as the outside world began to get an eyeful of what was going down in this place—the town had attempted to cut off the sightlines, erecting barricades, blocking paths, and obscuring the water with huge tarps, but there were still a few angles left. None of them were ideal or easy to get to, and often the monitors would show up at a formerly accessible spot only to find it suddenly "closed for repairs," cordoned off and under guard.

Despite the hazards and impediments, the monitors persevered, they were resourceful, and they always found a way. But their work was emotionally taxing. Watching hundreds of dolphins fight for their lives—and lose—was not a job for the fainthearted. Tim, a good-natured guy with a fullback's build, radiated steadiness, but the cove still took its toll. After witnessing an especially bloody day, Carrie told me, Tim would often be unable to talk.

Palmer had asked everyone to drop their stuff and then meet in his room to discuss tomorrow's schedule. Traipsing through the hotel, I had the paranoid sense of being watched through keyholes or by hidden cameras. When I arrived at Palmer's doorway I ducked in, relieved. Everyone was gathered around a Canadian man named Jack, who was pouring red wine into paper cups. "Alright," Palmer said, quieting us. Though he never sought the spotlight, Palmer was a natural at commanding a room. Tall and affable, usually clad in a baseball cap, he had an easy way with people and with language. At sixty-one, Palmer had spent his career in service to the environment. He had founded and led the Endangered Species Committee of California, lobbied in Washington, D.C., served as a Sierra Club chairman and vice president—for decades he had worked to save wild animals and wild places in countless ways, in hundreds of campaigns, in his own laid-back, humor-filled, get-it-done style.

"Let's begin with the subject of danger because it has been raised," he said. "What we usually do is throw Berman into the cove and if he makes it back, we know we're okay." People laughed, tension defused. Palmer continued: "Apparently there are extreme groups here—the police should keep them away from us. You don't want to go near them. By all means, do not get into a shoving match. There is always the possibility that you will be arrested along with the nationalist guy who was shoving you."

Getting arrested, everyone knew, was a particularly bad idea in Japan. You could be held for a month without being charged, and you would definitely be ejected from the country and not allowed to return. In 2007, a group that included the actress Hayden Panettiere had paddled into the cove on surfboards, placing themselves among a group of pilot whales the hunters were in the process of slaughtering. The protesters were jostled and screamed at and prodded with gaffs, and they exited the water quickly. It was a brave act and it drew some media attention, but they had to flee the country to avoid getting fined and detained. While it is naturally tempting to take an emotionally charged run at the cove, cutting the nets or otherwise interfering, the reason O'Barry doesn't advocate monkey-wrenching in Taiji is because it doesn't work. In his e-mails to the group, he stressed the importance of restrained, respectful behavior. Over the years he'd seen that haranguing the hunters just made them more determined to continue. Instead, he takes a different tact. He points out a jarring truth: that dolphin meat is about as palatable as industrial waste.

≈

It's a measure of Taiji's twisted reality that when they pass by this coast, dolphins might actually be helped by the fact that they are swimming Superfund sites, their bodies loaded with chemicals. Some of the worst pollutants bio-accumulate in dolphins' fatty flesh, including mercury, a potent neurotoxin. Mercury poisoning is no minor affliction: even low levels of it cause memory loss, nerve tremors, heart attacks, liver failure, loss of hair, teeth, and nails, blurred vision, impaired hearing, muscle weakness, high blood pressure, insomnia, and a hellish syndrome called desquamation, which is basically your skin peeling off. If you ingest even tiny doses of mercury, the toxin stays in your system and its effects compound over time.

When dolphin meat purchased in Taiji supermarkets—and served nationwide in schoolchildren's lunches—was proven to be awash in mercury in 2002, the Japanese media ignored this news. In 2008, after tests of Taiji residents showed sharply elevated mercury levels in their bodies, enough to cause brain damage and birth defects, again the press was silent. Alarmed, two town councilmen, Hisato Ryono and Junichiro Yamashita, paid out of their own pockets to print fliers informing people about the perils of eating dolphin. "This is a small town, where people are afraid to speak out," Yamashita said. "But we can't sit silent about a health problem like this." The party line about the dolphin hunt leaned heavily on cultural heritage, Ryono weighed in. "But they do it simply for profit. It's a business, not a tradition." (Having clashed with the political powers that be, Yamashita left Taiji, and I was told he now drives a cab near Tokyo. Under similar pressure, Ryono disavowed his statements.)

This issue isn't confined to Taiji. Though all dolphin meat and much whale meat is tainted—one sample was found to contain 5,000 times the limit for mercury contamination; another gave rats kidney failure after a single mouthful—the Japanese government has issued no warnings, other than advising children and pregnant women to eat it in moderation. "There is a real danger in whale and dolphin meat, but word is not getting out," researcher Tetsuya Endo from the University of Hokkaido, whose lab did many of the tests, told *The New York Times*. To another reporter inquiring about the edibility of dolphins, Endo snapped: "It's not food!"

Despite all this, the Taiji Fishermen's Union, the town's mayor, and the Japanese government continue to insist that the meat is fine. Tomorrow our main job was to contradict them, standing at the cove in our SAVE JAPAN DOLPHINS T-shirts holding signs that said in Japanese:

DOLPHIN MEAT IS CONTAMINATED WITH MERCURY. When Palmer finished explaining this—"Taiji is not just an animal rights issue; it's an issue of human rights"—an athletic-looking Asian man, Barry Louie, snorted in disgust. "It's pathetic," he said. "The government is poisoning their own people."

Louie lived in Osaka, but his family was from Hong Kong and he had grown up in California. He spoke fluent Japanese and was serving as a translator. Earlier, he had given us a primer on the country's etiquette: "There are lots of social rules. The most important one is to be nonconfrontational. In America, everything is in your face, right? You get in people's faces. And that's natural. It can even be funny. But in Japan, that's an absolute no-no."

"So let's head out and get some actual sleep," Palmer said, wrapping things up. "You should be down in the lobby at ten a.m., checked out with your baggage." The police had advised that we not leave the hotel before that, he told us, but then added: "I think it's okay if you're not alone. Stay in groups, stay in lighted areas. For those of you who want to go out, I'm happy to go with you—as long as I can outrun you."

≈≈

"YOU GO TO HELLLLL!!!!!"

The kid—or maybe he was an adult, I couldn't tell—stood in front of me with his fists clenched, screaming so hard that every tendon in his neck was visible. Spittle flew from his mouth. He wore a hat pulled low and wraparound sunglasses, a baggy black T-shirt, and pants at half-mast. He was pint-size and wiry and about as angry as any person I'd ever seen. With him were a few dozen of his friends. They were angry too; one guy was even holding up a sign that said, in bold block letters: ANGRY! They didn't seem to have gotten Louie's memo about avoiding confrontation. Behind them a squadron of police massed in riot formation, fingering their batons with crisp white gloves.

We had stepped off the bus into this crowd, the noise and fury hitting us like a wall. The day was hazy, damp with humidity and broiling with heat. For a moment, I felt dizzy.

"GET OUT OF HERE YOU TERRORIST!" another man yelled, leaning forward and punching the air with his fist; his malice was undercut by the cute Pomeranian he was holding on a leash. The tiny dog wagged its tail, excited by all the people.

"Go!" Louie shouted, moving us along. "Let's go!"

To get down to the cove we had to cross through a gauntlet of nationalists who had lined both sides of the road. They were shaking Japanese flags and berating us with varying degrees of English skill. "FAK RU! FAK RU!" one chubby, bespectacled guy shouted in an endless loop. The police surrounded us as we walked; the officers were professional and seemed genuinely interested in protecting us. Two vans drove by slowly with loudspeakers on their roofs, blaring more insults. On the street, our hecklers included a group of women dressed in traditional kimonos and porcelain-pale makeup. They looked lovely, from a block away. Up close, their faces were warped into sneers. "Why don't you die instead of the dolphins?" one woman hissed as we passed. Another woman took it up in a chant: "DIE! DIE! DIE! DIE! DIE!"

We made our way to the rocky beach, trailing policemen. The nationalists and assorted angry people stayed clustered on the roadway above. A pair of coastguard Zodiacs floated offshore, ready for potential skirmishes on the water. The men in them wore helmets and armored vests—a bit of overkill, I thought, given that our arsenal consisted of T-shirts and cardboard signs. Behind us the police fanned out in a long line. The sun flared through the clouds, bleaching the sky to white. It had to be at least 100 degrees out. A gaggle of cameramen and reporters were already down at the cove, and they surged toward us. "Smile and look happy!" Berman instructed.

Palmer, at the front of our group, stopped. "Okay. We're gonna be broadcasting from these rocks." One of the cameramen thrust a boom mike at him. "We'll have a series of events now," Palmer said, "starting first with a circle for the dolphins. A moment of silence—a prayer, if you will. For the dolphins who have died here in Taiji, and the dolphins who will die this year." He paused, leaving an extra-long beat. "We also wish to have a prayer for the people of Japan. Many died, as you know, during the earthquake and tsunami in March 2011. Today we honor the souls of the dolphins and the souls of the people."

Palmer's voice was warm but unemotional. It was the voice of reason, of hope that one day, instead of blind rage at the cove, there might be productive communication. Part of O'Barry's plan to end the dolphin hunt involved helping the fishermen develop ecotourism options. Taiji's coastline is stunning—the Wakayama prefecture includes nine World Heritage sites—and the cove itself is part of a national park that could plainly be put to better use than as a crude slaughterhouse.

Economically, dolphin watching would seem to make far better sense than dolphin hunting: there are fewer and fewer takers for the poison meat, and it sells for only six dollars a pound. But there is another financial incentive, and it has nothing to do with traditional whaling or ancient food customs or anything even remotely related to Japan's heritage. It has nothing to do, even, with the fishermen's stated goal of removing as many dolphins as possible from the ocean because they think the animals eat too many fish. Taiji's dirtiest secret—and the main reason its dolphin hunt continues—is that the town itself serves as a locus of live dolphin trafficking.

While a dead dolphin is worth maybe $500 to the Taiji Fishermen's Union, a healthy live dolphin—to be specific, a young female—can be sold for more than $150,000. In an average season about 10 percent of the dolphins driven into the cove are sold live, bringing in millions of dollars. In 2012, for instance, marine parks bought 156 bottlenoses,

49 spotted dolphins, 2 pilot whales, 14 Risso's dolphins, 2 striped dolphins, and 24 white-sided dolphins from Taiji. These dolphins were shipped to all corners of Japan, and also to Korea, China, Vietnam, Russia, and Ukraine, among other places.

Within a mile of where we were standing there were at least four dolphin-brokering businesses, each with its own trainers. After their capture the animals are deposited in holding pens and dingy concrete pools at one of these places, and taught some basic tricks, which increases their market value. As a result, Taiji is a one-stop shopping destination for anyone who would like to buy a dolphin, and who is untroubled by the process of plucking that dolphin out of a pool of blood that contains the dead bodies of its entire family.

We joined hands for our moment of silence, while above us people continued to yell with the full force of their diaphragms. The quieter we were, the more incensed they became. "FAK RU! FAK RU!" "GET OUT GET OUT GET OUT!" "YOU TERRORIST! YOU ARE TERRORIST!" One of the kimono women leaned over the roadside railing and let loose a fierce torrent of Japanese, sharp-edged with consonants. "What's she saying?" I asked Louie, standing nearby. "Oh, she's calling you a prostitute and she's asking: 'Shall I piss on you?'" he said, with a shrug.

"At least she's being polite about it," Tim said, overhearing.

Suddenly there was a scuffle at the edge of the beach. One of the nationalists had barged over to confront us, and the police had encircled him, slowing his progress. He was so upset he was literally shrieking, his head jutting frantically out of the huddle. It is hard to get riled, I noted, when somebody rips into you in a language you don't understand. It's like being attacked by marshmallows—the words hit but they don't really hurt. "He's saying to the police, 'You have no reason to protect the Americans,'" Louie translated. "'They know nothing of Japanese culture.'"

One of the triumphs of this year's group was that it contained a dozen Japanese members—a first-time, big-deal inclusion. Information about what was happening in the cove traveled slowly, the national media was once again muted, and local interests worked hard to quash any dissent. But despite ingrained beliefs against speaking up or standing out, within the country there was a blossoming homegrown movement to protect dolphins. One of the Japanese activists, a young man named Kai, stepped forward until he was toe-to-toe with the bellowing nationalist. Kai was thin and gentle, a quiet guy with a sweet smile. The nationalist outweighed him by at least eighty pounds. But when the two men faced off, Kai was unintimidated, staying centered while being hollered at, then giving it back twice as hard. After a while the police pried them apart and the nationalist stalked off, wiping his brow with a towel.

≈

Ric O'Barry arrived the next morning. We caught up with him in front of our second hotel, the Urashima, a series of white buildings wedged into craggy seaside cliffs. The hotel's architecture reminded me of a cruise ship, but it was more than a mile long. The ocean lapped on its doorstep. Toy ferries decorated to look like dolphins and whales—equipped with tail fins and googly cartoon eyes—transported guests from the lobby to the main streets of Kii-Katsuura.

Though he had flown for thirteen hours, endured his usual three- or four-hour grilling by officials in the Tokyo airport, and then driven seven hours to get here, O'Barry took the time to give a press conference before he'd even checked into his room. He spent an hour speaking to some Japanese reporters, standing on the street in a blue hoodie, gray hiking pants, and flip-flops, hiding tired eyes behind Ray-Ban sunglasses. His white hair strayed out from under a khaki baseball cap.

"What do you . . . um . . . what is your wish, what do you want the fishermen in Taiji . . ." asked an earnest-looking man, holding a notebook.

"To stop killing the dolphins," O'Barry said, with exasperation. "It's really simple."

No matter how sleep-deprived, O'Barry is a lively interview subject and a formidable debater. He is plainspoken and direct and he delivers his message in calm, deliberate tones, with frequent jabs of wry humor. He is a master at pointing out absurdities and hypocrisies in the realm of marine parks: "We love our dolphins like they're our family—I hear that a lot. Really? You lock your family in a room and force them to do tricks before they eat their dinner?"

The reporters drifted away and O'Barry sighed, jamming his hands in his pockets. He looked exhausted and not thrilled to be back at the cove for a tenth straight year. "Basically I would like to be put out of business," he told me. "That's my goal. But it won't happen in my lifetime. There's too much money in captivity." O'Barry had firsthand experience with these stakes: he'd flown here directly from a Florida courtroom where he was one of the defendants in a battery of lawsuits involving twelve dolphins purchased in Taiji. He was being sued for defamation and tortious interference; the plaintiff was Ocean World, the marine park I'd visited in the Dominican Republic. If you toted up every last charge, Ocean World was demanding damages in the billions.

The dustup started in 2006 when O'Barry, hidden in a lookout above the cove, watched a group of bottlenose dolphins being picked out of a corralled pod. The fishermen killed the rest. When O'Barry discovered the dolphins had been contracted for—that Ocean World had ordered them through a Taiji dolphin broker and hired two American veterinarians, Ted Hammond and Michael B. Briggs, to oversee the transfer—he made that information public, along with his eyewitness account of the animals' violent capture. The ensuing outcry was piercing and the Dominican Republic government, fearing loss of tourist dollars, denied the dolphins' import permits. Ocean World was thus left with a dozen dolphins for whom it had paid $154,000 each, but couldn't take home from Taiji.

In a case filing, I would later read Ocean World's lawyers' interpretation of the events at the cove: "The plaintiff, Ocean World . . . made a reasonable effort to help the Twelve (12) dolphins embark in a second chance for life and to assist them in a journey involving safe passage to a loving caring home at his family's one hundred million dollar $100,000,000 world class facility." The document wound on for thirty typo-riddled pages, alleging that O'Barry's actions were part of a grand self-promotional scheme to sell more books and DVDs, while interfering with Ocean World's noble and charitable attempts to adopt the dolphins.

O'Barry was represented by prominent lawyers who took the case pro bono because they deemed it a SLAPP lawsuit (strategic lawsuit against public participation), charges intended to squelch free speech. The goal of a SLAPP is not to win in court, but rather to exhaust the defendants in every possible way (especially financially) by filing endless motions, charges, deferrals, requests for documents, subpoenas for depositions—a blizzard of legal paperwork. By now, more than six years later, Ocean World had filed another lawsuit against O'Barry (which was dismissed); sued two scientists who had also spoken out against the Taiji captures (and the universities where they worked); the original complaints had sprawled to twenty-eight file boxes, and there was still no end in sight.

≋

Berman and I wanted to visit the infamous Taiji Whale Museum, located next to the cove. The museum's name is misleading: it is actually a *whaling* museum. Along with exhibits on whale and dolphin hunting, it also displays dolphin species that are rarely seen in captivity because they tend to expire in short order. I mentioned to O'Barry that we were headed there and he nodded resignedly. "I hate that place," he said. "It's the Bates Motel for dolphins."

While O'Barry restored himself with a nap, Berman and I cabbed it back to Taiji, getting out near a traffic roundabout decorated with a pair of life-size whale tails bursting from the ground. Masako Maxwell had also joined us, a Japanese American technologist from Los Angeles who devoted herself to helping animals of every species. She had a ponytail down to her waist and a tattoo ringing her biceps and she carried herself with the quiet strength that is the sure mark of a badass. Maxwell didn't have to make a lot of noise—she just got things done. "I was born and raised in Japan and I feel it's my mission to come here and be useful," she'd said, during introductions. "She's being very modest, of course," Palmer had interjected. Maxwell, he said, ran the group's Japanese Web site and all of its social media. "She's key to getting our information out to the people of Japan."

For Berman and me, Maxwell was also key to our chances of getting into the building. Far from being a neutral educational or scientific facility, removed from the controversy and carnage that was going on at the cove, the Taiji Whale Museum was one of the town's central dolphin traffickers. Alongside tanks that housed its performing dolphins were additional floating pens that contained animals for sale. Although it is a public building, its ticket windows were covered with signs that declared: NO ANTI-WHALERS ARE ALLOWED INSIDE THE MUSEUM.

A reasonable person might wonder how the cashiers would make this determination. Was there a secret handshake, known only to fans of the dolphin hunt? In the end, the Taiji Whale Museum had decided to cut right to the chase and simply refuse admission to any Westerner. Past experience had shown, however, that once a Westerner did get his hands on a ticket he could usually get past the door—unless he exhibited signs of being less than completely enamored by an establishment where, as O'Barry liked to point out, you could buy dolphin-meat snacks and eat them while you watched the dolphin show.

Maxwell went to get the tickets. Berman and I lurked out of sight, hiding ourselves behind a hedge. I looked down and noticed that I was standing on a ceramic tile painted with dolphins. Earlier, I'd walked across metal plaques sunk into the pavement, etched with splashing whales and dolphins and the words WELCOME TO TAIJI.

Everywhere you turned in this town, there were cetaceans. Whales were plastered on buildings; dolphins were pretzeled into neon signs. They ushered you to the bathroom and saluted you at storefronts and pointed you down streets. I don't think I have ever been surrounded by as many finned creatures as I was in Taiji, an irony that was difficult to process. WE LOVE DOLPHINS! a prominent road sign exclaimed. It was like being in the Twilight Zone, with flippers. It didn't seem like the town could get any more schizophrenic, but then Maxwell appeared with the tickets, we pulled on our hats and sunglasses and made our way into the museum, and I realized that every bizarre experience I'd had until now was only a warm-up for this.

Inside, four whale skeletons hung from the ceiling, enormous mobiles of bone. A model of a live whale dangled up there too, pursued by a boat filled with a dozen men hurling spears. Below, there was a puppet show depicting how the animals were killed: you could press a button and watch the fleet attack a whale that popped up from a hole. Harpoons of all shapes and sizes and vintages were on view, along with maps of celebrated whale-hunting grounds. But if the first floor was a history lesson, the second floor was all about biology.

The first thing I saw at the top of the stairs was a glass case that contained the head of a striped dolphin. The head was pinkish, suspended in pale yellow fluid. Its eyes were open, which made it disturbingly real, two unseeing orbs staring out for eternity. The dolphin was smiling, as all dolphins do, proving with finality that this feature of their anatomy does not mean they are always happy. Lined up near the head were four cylinders filled with dolphin fetuses in various stages of development. They were squashed in, so their tails curled under them and their beaks tilted skyward. I was startled to see that their fledgling fins looked exactly like arms. Like us, their heads are large throughout their gestation, so there were odd echoes of humanity in the bobbing creatures, even with their comma-shaped bodies and pointy faces.

The whole floor was a gallery of specimens: a floating bottlenose brain, a pickled humpback whale embryo, tissuey slices of . . . something. Maxwell walked down the row, reading off the contents of glass bottles and jars: "Whale penis, dolphin penis, whale heart, whale tongue, whale anus, whale spleen. And, oh! This one is an orca."

It was hard not to be shocked by the sight of a stillborn killer whale lying on its side in a liquid-filled case with its ropey umbilical cord still attached. You could just barely make out its coloration, a whisper of difference where black met white. The orca's small body was so smooth it glistened, as though it were made of pristine, flesh-colored custard. Looking at it, Berman let out a long breath, shook his head, and walked away.

We retreated downstairs and through the gift shop, past the freezer filled with dolphin and whale meat, the cans of dolphin stew stacked next to the dolphin plush toys and T-shirts and key chains. Outside, swelling muzak signaled the start of the dolphin show. We took our seats at the top of the bleachers so we could watch the audience as well as the show. Behind us, a blue whale skeleton hovered like a spaceship, its back end hoisted as though it were diving. It was hardly a packed house, mostly parents with children. The day was intolerably muggy, and a few toddlers were screaming in the heat. Mothers fanned themselves; fathers looked stoic and bored. Berman and I were the only smuggled-in Westerners in the joint.

A squad of six uniformed trainers took up their positions—young women in tangerine polo shirts and navy Bermuda shorts—and the show began in an area like the cove, except this one was walled off from the open ocean by a bulky cement barrier. There was no tidal flow in here, no exchange of fresh seawater with the waves, no wayward fish. The enclosure was still and stagnant and hot. One of the trainers blew her whistle and a Risso's dolphin erupted from the water, followed by a pilot whale who had to be fifteen feet long. A third oversize dolphin flipped onto his back and began to swim by us, waving his pectoral fins. Berman looked deflated. "That's a false killer whale," he said. "A very pelagic deep water animal. He won't last long here."

I was still stuck on the Risso's dolphin. He was the most unusual dolphin I'd ever seen—a Cy Twombly dolphin, his gray-blue body covered in fantastic scribbles. He looked like an adorable alien. Actually, they all did. The pilot whale's jet-black head was almost perfectly round; the false killer whale still bore traces of the dolphin's earliest incarnation as a sleek, wolfish creature. All three dolphins were magnificent, absolute marvels of the ocean, and by all rights they should have been out in the Pacific, doing what 55 million years of evolution had designed them to do in the most important ecosystem on earth, instead of in here, leaping to the beat of cheesy pop songs.

As I watched, sweat trickled down the back of my neck, but something else was rising: anger. The show was soul-crushingly stupid. It was plainly and inanely stupid—all of this was stupid, everything that went on at the cove, the entire arrogant, selfish relationship we had with these animals and with all of nature, as though every bit of life existed only for our purposes. We behaved as though we were gods, deciding the fate of everything, but we weren't. We were just dumb. I felt a wave of despair wash over me.

Berman slumped forward. "These trainers," he said. "How can they possibly live with themselves? They treat these animals as pets but they watch while the others are slaughtered. It's mind-boggling." He wasn't alone in this sentiment. Over his years observing the cove, O'Barry had seen trainers wading in to grab dolphins by their tails, trainers riding in dolphin-hunting boats, trainers colluding in every way. One time he had even seen a trainer point out an escaping pilot whale so the hunters could recapture it.

The performance ended. I was in a dark funk and would have loved to leave, but Berman wanted to check on the bottlenoses in the indoor tanks. They were in a circular white building at the far end of the dolphin pens. Inside, it reeked of chlorine. There was a stuffy, listless quality to the air, as though its main goal were suffocation. Three bottlenoses were crammed in a shallow tank that arched over a walkway, its windows dirty and cloudy and scratched. One dolphin swam up to Berman and hung in front of him, looking him directly in the eye. Berman touched his hand to the plexiglass. "You want to go home, don't you, buddy?" he said softly.

The walkway ended in a murky aquarium, lit by buzzing fluorescent tubes. One shoebox of a tank contained three spotted porcupine fish, a species I especially love, and as I watched them fluttering hopelessly my mood sunk further. No effort had been made to include coral or any kind of ocean features; an electrical cord encased in plastic was the tank's only décor. The overall effect was of a fifties-era mental hospital for fish.

Depressed, we headed for the exit, and passed a trainer feeding the false killer whale and the pilot whale. The animals spy-hopped in front of her, their heads out of the water and their mouths open. The trainer had short hair and a cheery round face. She had barely cleared her teens. "Hello," she said.

"Oh, do you speak English?" Berman said, stopping abruptly. "Can I ask you a few questions?"

The girl examined Maxwell warily and replied in Japanese.

"She won't let you take video," Maxwell reported, "but you can ask her some questions."

"You know they hunt dolphins here?" Berman asked, wasting no time.

The girl paused. Her face suddenly looked less cherubic. "Yeah," she said.

Berman looked at her. "So do you feel sympathy? For those dolphins?"

The girl stared back at him. She puffed out her cheeks and crinkled her nose. "Mmmmmmmmmmm," she said, moving the air from one cheek to the other. "Mmmmmmmmmmm." She seemed to do this for about ten minutes, deliberating. "Sympathy?" she said finally, then spat it out sharply: "NO."

Maxwell and I glanced at one another. A security guard, noticing the conversation, was walking briskly toward us.

"I'm just asking your personal opinion," Berman pressed. "Not the opinion of this place."

The girl inflated her cheeks again. "I'm not feeling sympathy because sometimes people hunting deer and they are hunting . . . cow or something. And I can't recognize what's the difference." She pointed to the false killer whale and the pilot whale, who were staring so intently at us from such a close distance that they seemed to be part of the conversation. "I know it's really intelligent," she said, with a shrug. "But I feel that cows are also really intelligent and we are willing to eat them. And the number of dolphin species are increasing, so . . ."

The security guard was upon us. He was a beefy dude and he didn't seem pleased. "Whale-as! Whale-as!" he barked, making a shooing gesture. Rather than argue with him, we left.

It wasn't the first time I'd heard someone defend the dolphin hunt by accusing others of similar mistreatment of animals. But if the point of the hunt is subsistence, then simultaneously selling the animals for six-figure prices is impossible to justify. If dolphins are extremely valuable, then how can they also be completely dispensable? Not to mention that when we kill any creature for food we have ethical obligations: to do a clean, swift job of it, to avoid taking endangered species, to show respect and gratitude always, to tread as lightly as possible on the balance of life in an environment. None of these things were happening in Taiji.

≋

O'Barry and I met in the hotel lobby that afternoon. The Urashima's ground floor reminded me of an airport terminal, if every traveler who passed through it were wearing a *yukata*—a kimono-like cotton bathrobe—and rubber flip-flops. There were at least ten *onsens*—hot spring baths—sprinkled throughout the property, so many buildings and wings and tunnels and corridors that if you weren't armed with a map you could be lost for days. As a helpful guide, the hotel had painted colored lines on the floor: take green for the cave baths, orange for the lava-rock baths, yellow for the shrines, red for the buffet halls.

I hadn't eaten in a while—the combination of stress, heat, and murderous conflict is not very appetizing—so we decided to have lunch at a restaurant O'Barry liked in Kii-Katsuura. As we crossed the street, a car caught my eye. It was a black Infinity sedan, parked between two nondescript buildings. Yesterday I'd seen it cruising the cove. The men in the car were yakuza, Palmer had pointed out, part of Japan's organized crime. They tended to insert themselves into lucrative industries, and dolphin trafficking apparently qualified. I had noticed the Infinity as we'd exited the bus—it stood out among the subcompacts and battered loudspeaker vans—and I'd gotten a close look at the men inside it. They were dressed differently than the nationalists: instead of rising sun T-shirts and polyester track pants, they wore hipster sunglasses and understated dark clothes. They all had shaved heads. Even the police had avoided them.

"Hey," I said, nudging O'Barry. "I think those guys were at the cove yesterday." The two men sat in the front seats eyeing us coldly.

When you describe someone as "a person you wouldn't want to meet in a dark alley," these were the type of people you meant—and here they were right now, in a dark alley.

O'Barry turned to look. "I know him," he said, pointing to the man in the passenger seat. "Last year that guy threatened to kill me on camera. I'll send you the footage. He is screaming into the camera, 'I'm gonna kill you, O'Barry! I kill you!' Oh yeah, he's a yakuza."

"He seems pretty scary," I said.

"He's a nut-job!" O'Barry studied the car. "That's why I'm a little afraid of him, because people who aren't rational—you know, they have a little too much sake and anything could happen."

I was disturbed by the fact of unhinged criminals having tailed us to our hotel, but O'Barry seemed to take it in stride. For him, being menaced was part of the job. "If you can get a dolphin in the right place, you can make a million dollars a year off that one dolphin," he had calculated, a fact that exposed him to all kinds of dangers when he showed up and proposed to set that dolphin free.

Not long ago in Jakarta, O'Barry had been advised by police to wear a bulletproof vest after successfully lobbying to shut down a traveling dolphin show—a vile production in an underground parking garage that featured dolphins jumping through hoops of fire. Later during that same trip a gang of thugs interrupted a talk he was giving, hosted by the U.S. embassy. Also, he had awoken in the night to the sounds of someone trying to break open his hotel room door. While it was never easy to pry dolphins away from people who profited from them, some situations were more perilous than others. O'Barry had recently been told about the desperate plight of two dolphins in the mountains of Turkey, and he planned to go there soon to see what he could do. "But it's going to be very difficult," he told me, "because the owner is part of the Russian mafia, and he has the dolphins in his swimming pool."

O'Barry is involved in so many dolphin affairs, so many protests and rescues and initiatives, that it is hard to keep track of them. He has come to dolphins' aid in the Bahamas, Mexico, Nicaragua, Guatemala, Panama, Colombia, Haiti, Indonesia, Spain, Switzerland, Germany, Singapore, Britain, Egypt, Israel, China, Canada, and the Faroe Islands, among other locales. "I never planned on being an activist," O'Barry said. "But one thing leads to another. Now, if there's a dolphin in trouble anywhere on this planet my phone will ring."

One of the hairiest stops on O'Barry's circuit is the Solomon Islands, an archipelago just east of New Guinea, one of the poorest countries on earth and also one of the roughest. In rural communities, which comprise most of the place, dolphin teeth are used as currency. Because of this, dolphin hunting is practiced in even the smallest villages; dolphin trafficking has also sprung up, with turbulent consequences. "There are problems all over that place," O'Barry told me, his voice freighted. "Life is very cheap there." Earth Island's Solomon Islands director, a man named Lawrence Makili, had been beaten within an inch of his life. Makili, who happens to be north of six feet and 200 pounds, fought back and managed to escape, though with terrible injuries. Two other colleagues of O'Barry's had not been so lucky. They were both murdered while trying to stop dolphin trafficking. Jane Tipson, an outspoken activist in St. Lucia, was shot in the face at point-blank range; in Israel, another, Jenny May, was strangled with her own belt. No one was arrested for either crime.

We boarded the dolphin ferry and set off across the bay. Kii-Katsuura was a mellower town than Taiji, bigger and more sophisticated, and in the businesses and streets there was less free-floating animosity. Still, I noticed a lack of enthusiasm about our presence, store clerks suddenly becoming very preoccupied when we approached, turning their backs or vanishing entirely.

For O'Barry the distaste wasn't mutual. He detested the doings at the cove, but after spending so much time here he'd come to appreciate everything else. "This town reminds me of Miami in the fifties," he said. As we walked, he pointed out a bakery he visited each morning, and a toy shop that didn't bother to lock up at night: "There isn't even a door!" He nodded admiringly. "There's a lot that's right about this place and these people."

O'Barry stopped in front of a restaurant that displayed models of its menu items in the windows, molded out of plastic. "I usually pick what I want here," he said, gesturing at the shellacked dishes. We went in and sat at a banquette. When the waitress approached, O'Barry greeted her warmly in Japanese. In the background, unlikely accordion music was playing. After she took our orders I filled O'Barry in on our trip to the Whale Museum, an institution he battled constantly. On one tense occasion its manager, a trim businessman named Hiromitsu Nambu, had waved a samurai sword at him. The two men were long-time nemeses. Upset that the museum's dolphins suffered from blistering sunburn, O'Barry had asked if he could pay for an awning to shade the outdoor tanks. Nambu had agreed to let him. "That was six years ago and he *still* hasn't done it," O'Barry said in a sardonic tone, leaning on words for emphasis. "And he's not *going* to do it because he doesn't *care* about dolphins. And he's a *Buddhist*!" He rolled his eyes. "Theoretically."

"Have you ever seen a Risso's dolphin?" I asked, still fixated. "They're incredible-looking creatures."

O'Barry nodded wearily. "Yeah," he said. "They kill a lot of them here."

Sometimes when O'Barry talks about the cove, he just seems tired. Tired of fighting, tired of watching dolphins die, tired of the media's short attention span, tired of schlepping here. But at other times something unusual happens: his whole persona both hardens and melts, in the manner of an expert martial artist. His eyes become intense but his body remains relaxed, and all of his energy streams into the moment. He isn't fearless—that would be silly—but he is ready, in a quietly defiant way, to face his opposition. It is a resilience, I figured, that he'd developed over time, the way you'd build a muscle. In Taiji, the town was malefic and the people could be horrid, but the cove's most demanding challenges were personal ones: How do you survive your own sadness?

"What's it like when a bunch of dolphins are in there?" I asked. O'Barry looked down and rubbed his hands together. I noticed the dolphin tattoo near his left thumb, its edges faded by saltwater and time. "It's heartbreaking," he said, from somewhere deep in his chest. "Because you *know* what's going to happen. I've seen as many as three hundred dolphins in there—pilot whales, false killer whales, bottlenose—all in one day! Yeah, when you're actually seeing it up close and personal it's much more . . . it's not like watching it in a movie. You can hear them, and at a certain angle you can sometimes see them throwing themselves onto the rocks, trying to escape." He paused again, struggling for words. "You know, it's . . . anguishing. 'Anguish' is the one where you can't do anything."

Spelunking daily into the depths of grief and heartbreak is not something most people want to do. Over the years, interviewers have asked O'Barry why he chose to devote his life to dolphins in such a single-minded way, setting them firmly on the front burner and everything else, including his family, on the back. When you get to know him, however, his motivation becomes obvious—he has no choice. O'Barry understands dolphins and what they represent: "They're a reference point in our relationship with nature. The cove opens to a bay, the bay opens to the sea, life goes on." As in a holograph, what happens in one little corner of Japan contains the blueprint of the whole. Domination, cruelty, profiteering: what a tragedy if we let those actions define us.

≈≈

I found the dolphin and whale sushi bar in one of the mildewy tunnels at our hotel, an unassuming restaurant cloaked by a canvas curtain. The only thing that signaled its purpose was a whale on the front of the menu, swimming across a soup bowl. I poked my head in and saw that the place was full. Near the doorway, a Japanese family, cozy in their *yukatas,* sat at a table laden with dishes and plates and kettles, bowls of broth, and bottles of beer.

I'd come upon the restaurant while trying to return to my room after visiting the Urashima's most popular *onsen* bath, the Boki-do Cave, a volcanic pool on the windward side of the bay. The cave was blustery and wave-swept and dramatic, all things I liked, but I had left quickly. There is a list of *onsen* rules about three feet long, setting out many compulsory and forbidden behaviors, and though I'd been instructed about when to rinse and what never to do with soap and when you must be naked and when it is unspeakably rude to be unclothed—I'd forgotten everything, distracted because every person I met was glowering at me. I hadn't even been sure about whether I'd stripped down in the women's changing room or a coed resting area; both were identified only by gold Japanese characters on red velvet drapes. I took a wild guess and, judging from the looks I got, wasn't sure it had been the right one.

Touring around in my *yukata,* feet slip-slapping in my shower shoes through the maze of hallways, I opened a pagoda-shaped door and stumbled onto a banquet hall—private dinner for two hundred bathrobed guests in progress, no one overly pleased to see me gaping at the buffet's centerpiece: a three-tiered sunburst made of tuna heads, surrounded by platters of raw tuna, and every tuna's mouth stuffed with yet more hunks of tuna.

It was time to leave. I was tired of looking over my shoulder for malingering yakuza, and the weather had turned so foul that the dolphin-hunting boats were confined to shore. Maxwell was driving back to Osaka tomorrow, and I was going with her.

We had agreed to meet at four o'clock the next morning, rising in the dark so I could catch a noon flight from Osaka back to New York

City. When I rolled my luggage across the lobby in the predawn, the cavernous hotel seemed lonely without the crowds rivering through it. Outside, the sky was a black hole, rainy, moonless, and illuminated only by the wet glare of dock lights. While I waited for Maxwell, I texted O'Barry to say good-bye. Last night he had taken the monitors out for a vegan Chinese dinner and I had missed the chance to do it in person. "Good luck at the cove," I wrote. "I hope the bad weather holds up." I asked if I could stay in touch, maybe catch up with him at another stop on his unending itinerary. I wasn't surprised when he replied immediately. O'Barry rarely slept while he was here; that kind of peace eluded him. He wasn't sure where he was headed after Taiji, he texted, but he would let me know. His travels might take him to the Philippines or back to Indonesia or hopefully to Denmark, where his wife, Helene, and his eight-year-old daughter, Mai Li, lived. He hadn't seen them in weeks. There was a dolphin protest in Ontario, and a baby dolphin at loose ends in Spain, and one particularly tough trip that he would need to make soon, to the epicenter of dolphin turmoil. If I didn't mind a level of hazard and confrontation that would make Taiji look restful, I was welcome to join him, O'Barry wrote. "I still have a lot of unfinished work in the Solomon Islands."

CHAPTER 6

A SENSE OF SELF

It was a long straight shot from Vegas to Utah, driving from the neon blitz of the Strip across miles of blast-furnace desert, studded with brush and aching for water, to Mesquite, where the landscape turned as red as Mars. I stopped there for the night at a motor hotel perched above the town. From my room I could see miles of canyonlands, red rock mesas banded with mauve and rust and pink glowing under a fat moon. Early the next morning I drove on, leaving Utah for a second or two to clip Arizona's northwest corner, rolling past creepy Colorado City, where members of a polygamous sect of Fundamentalist Mormons lived with their child brides and fifty-eight-member families in oversize boxy houses. The road was flat and featureless and invited speeding.

Back into Utah again, just past the state border, I turned east toward my destination: Kanab, population 4,410. That number had recently increased by one. An eminent neuroscientist, Lori Marino, had packed up her lab at Emory University in Atlanta, said good-bye to her collection of dolphin brains, and moved here. It wasn't Kanab itself that had drawn Marino, but one particular piece of it, Angel Canyon, where the Best Friends Animal Society runs the country's largest no-kill sanctuary, housing some two thousand dogs, cats, horses, birds, rabbits, and other lucky creatures. On Best Friends' 20,700-acre grounds, Marino's neighbors would be one-eyed bunnies and rehabilitated eagles and abandoned piglets and neurotic donkeys and battalions of kittens and some of Michael Vick's pit bulls, a stellar, crazy-quilt community of animals. Which was exactly what she wanted.

I had become interested in Marino after reading her name constantly in relation to dolphins. She specialized in biopsychology, the biological basis of behavior, and neuroanatomy, the study of the brain's architecture, but instead of focusing her attention on humans, Marino had chosen to study cetaceans. She was one of only a few dolphin brain experts in the world. When dolphins appeared in the media she was often quoted, and what she had to say was always engaging. In one interview I'd read, she was asked: If dolphins had stayed on land, would they now be the dominant intelligence on Earth? I suspected Marino did think that, but she answered in a nuanced way. "While they don't build rockets, their level of sociality is so sophisticated I don't think they have anything to learn from us," she replied. "The fact that they have co-habited the ocean and not destroyed themselves really speaks to the fact that they have figured out a way to do this in a way we haven't."

Marino's work was very prominent in the scientific literature: she had authored more than a hundred papers in her field. "The dolphin brain represents a different neurological scheme for intelligence," she said, explaining her research. "And it's a very complex intelligence." In one massive study, she had collected 210 dolphin skulls from across the eons and run them through CT scanners to determine how the animals' brains had evolved. Marino and her colleagues reconstructed 3-D models of dolphin brains from as far back as 47 million years ago, when they were relatively small and unspectacular, and charted them up to their current, turbocharged state.

Curiously, as the dolphins' brains expanded dramatically in size, their bodies shrank, their teeth became smaller, and they developed high-frequency hearing. This downsizing contrasts with our own far more recent developments—as early human brains grew, during our own surge between 800,000 and 200,000 years ago, our bodies did not significantly change scale. Our sensory abilities stayed more or less the same. We didn't suddenly turn into dwarves or sprout wings or learn to see through our noses. It was business as usual for us, except with more horsepower, but the dolphins have shape-shifted in bold ways. On several occasions during their 95-million-year existence they have morphed into entirely different creatures, adapting to life both on land and in the ocean, appearing at various points as solo predators with impressive fangs, ace communicators packing powerful sonar, social networkers juggling complex relationships. Their bodies have gone through constant flux. But what happened during the dolphins' history, Marino wondered, for their gray matter to undergo such a radical growth spurt? It was an evolutionary puzzle. I wanted to talk to her about this, and countless other dolphin perplexities. She had looked into the animals' heads, literally.

Marino had also investigated more conceptual aspects of the dolphin mind. In 2000, she and another scientist, Diana Reiss, conducted one of the most acclaimed dolphin experiments of all time. Its premise was deceptively simple: a test to see if a bottlenose could recognize himself in a mirror. Most animals couldn't. They ignored the mirror or treated it as if it were another animal, approaching it tentatively or aggressively. To make the test more definitive, the subjects' bodies were marked in a conspicuous spot. For instance, if a chimp who was daubed with a pink stripe down his cheek leaned in to touch his mark and examine it in the mirror, he was said to have "passed" the test. He understood that the chimp with the odd tattoo was himself.

When Marino and Reiss first proposed to try this with dolphins, only humans and our fellow great apes—chimpanzees, orangutans, and gorillas—had demonstrated self-awareness. So it was front-page science news when Marino and Reiss's two bottlenoses, Presley and Tab, became the first non-primates to do this, mugging in front of the mirror, craning their bodies around, and flipping upside down to examine their marks. (Since then, elephants and magpies have also passed the test.)

Though it might seem like no big deal, to conceive of your own identity is a rare cognitive feat. The idea of a *self* is a pretty far-out abstraction, and to get that I am me and you are you and that we both have autonomy but there's also a relationship between us—this capacity was long considered unique to our own two-legged, opposable-thumbed species. It's not an ability that can be taken for granted: children don't begin to develop self-awareness until they're nearly two years old, along with feelings like sympathy and empathy. To know that dolphins operate in the same realms of consciousness we do raises a raft of fascinating questions about their interior lives and, in turn, the ethics of how we treat them. What Reiss and Marino accomplished, really, was to prove what John Lilly had only been able to guess: that the dolphin in the tank is not a *what* but a *who*.

In the aftermath of the mirror test, Marino did something unexpected: she publicly vowed never to conduct research on captive dolphins again. Most scientists tack in the opposite direction, leaning away from their feelings in an attempt to focus on the data; for Marino, it was the data themselves that prompted her to take a stand. Knowing how conscious the dolphins were of their own situations, she could no longer justify keeping them in tanks, away from their pods and their natural lives. These days she refers to herself as a scientist-advocate, using everything she's learned about dolphins to argue for their well-being.

Mixing hard-nosed science with heartfelt activism is a delicate balancing act, one that might trip up the career of a lesser scientist, but Marino's credentials make it impossible for anyone to dismiss her. For instance, if you are someone who would like to tell the public, as SeaWorld does on its Web site, that dolphins' high intelligence is "untested and disputed," Marino would beg to differ with you. Then, she would like to point you toward several dozen peer-reviewed studies that prove you wrong. Testifying in Congress in 2010, she demonstrated that marine parks violate the U.S. Marine Mammal Protection Act, which requires facilities displaying dolphins to also provide accurate educational materials. Marino examined the parks' offerings line by line and pointed out multiple errors, misrepresentations, and even lies, shredding them to bits by citing the actual science.

It wasn't only dolphins Marino wanted to stick up for. By moving to Angel Canyon she was signaling a next phase in her career, one in which she would use hard facts to petition for all animals. She wasn't alone. Researchers around the globe are coming to the same conclusion—we are not the only beings who matter—and new ideas are stirring about how the startling depth and breadth of other creatures morally obliges us to act humanely toward them. Now we have seen that elephants cry when they're sad, and some dogs have a greater vocabulary uptake than toddlers, and sheep can pick faces out of a crowd. We've learned that chickens show empathy and pigs express optimism. Scrub jays plan for the future. Hippos can be vindictive. Pigeons are excellent at math. Thanks to YouTube, viral video clips of animals doing amazing stuff are a regular feature in our lives, cats rescuing their owners and rats cuddling stuffed toys and bonobos driving golf carts. But what will we do with this information?

In 2012, a group of neuroscientists at the University of Cambridge drafted "The Cambridge Declaration on Consciousness," which recognized the awesome abilities of nonhumans, down to the lowly earthworm (makes judgments, acts with discernment, strategizes). "The body of scientific evidence is increasingly showing that most animals are conscious in the same way we are, and it's no longer something we can ignore," one observer wrote of the proceedings.

Marino wanted to do more than endorse this declaration; she wanted to act on it. Throughout her career she had watched fellow scientists perform horrific animal experiments, tinkering with their subjects in Frankensteinian ways and considering it standard practice. During her PhD, she had turned down a full scholarship at Princeton because she couldn't bear to vivisect cats. As a student, Marino's work with lab rats had given her nightmares; years spent studying brains taught her beyond all doubt that there are no dumb beasts. Despite the seventeenth-century rationalist philosopher René Descartes's proclamation that animals lack a soul—that they are no more than sentient machines—we are now aware of a far more textured reality: that other brains have bloomed along with ours, forging many divergent paths to intelligence, none of which we fully understand. Dolphins, everyone agrees, are Exhibit A.

≈

Kanab is a tiny town on the edge of a vast wilderness, with the Grand Canyon only three hours up the road. It has a folksy feel. I was relieved to be out in fresh and appealing scenery, in a place where people were more bent on rock climbing than dolphin killing. Back in New York I'd noticed that Taiji was stuck to me like tar. It made me feel out of sorts, even nauseous sometimes. The nerve-racking clamor of Manhattan—the pounding jackhammers and shrieking car alarms and ceaseless traffic, the sirens, street cleaners, clanging noises and small explosions emanating from trucks—seemed to rattle me more than ever. There were too many crowds and too much cement and not enough trees. Sleep was elusive, punctuated by dark dreams. One afternoon on Forty-second Street a heavily pierced girl elbowed past me on the sidewalk wearing a T-shirt emblazoned with the words FUCK EVERYBODY. I understood how she felt.

I couldn't get the dolphin hunt out of my head, especially since each day brought new bad news. Two days after I'd left, the hunters had captured a hundred pilot whales and killed every last one of them, including mothers and calves, along with a pod of thirty bottlenose dolphins. Tim Burns, monitoring the cove, had written, "I had to re-count numerous times to actually wrap my brain around such an alarming number. I'm speechless." The next day he reported: "Dolphins giving up a fight. Fishermen ramming them with boats and motors. Ruthless." Rumors circulated that the Taiji Fishermen's Union had run out of freezer space and was now gathering dolphins for live export. A pod of ninety-two bottlenoses was corralled next, and the marine park selection process lasted for days. Most of the dolphins were sold, and the remainder—too old, too young, too scarred-up, or too feisty—were butchered as usual. Then a group of Risso's dolphins was driven into the cove. None of them made it back out. "Things are bloody awful here," O'Barry had e-mailed.

Hitting the road has always seemed to me like a fine antidote for angst, so I had been itching to head out. It was only after driving hours into the desert that I felt myself unclenching. There is no ocean in Kanab, but there is the peacefulness that arises when nature is bigger than you are. Five miles outside town, I spotted the Best Friends sign and turned into Angel Canyon.

Marino was waiting for me at the visitors' center, a ranch-style building at the foot of some red rock bluffs. She is petite, with shoulder-length sandy-brown hair and large, watchful hazel eyes. We sat outside on a deck, while beside us hummingbirds whirred around a feeder. Thunder rumbled in the distance. I told Marino that I was happy to be somewhere serene after grappling with Taiji. She nodded, looking pained.

Marino hadn't been to the cove, but she'd seen plenty of footage.

Years before *The Cove* hit theaters, before the fishermen began their efforts to conduct their slaughter out of sight, filmmaker and activist Hardy Jones had captured graphic video of the hunts and distributed it widely. The brutality of the process shocked every marine researcher who watched it. Though scientists rarely agree unanimously on anything, more than three hundred of them signed a letter to the Japanese government denouncing the dolphin hunts. Marino helped organize this effort, and made her own outrage clear. As a result, she (and Emory University) had also been sued by Ocean World for speaking out against the capture of the Taiji twelve. Like O'Barry, Marino was litigated against for millions of dollars. Her case had been settled so she couldn't discuss details, but she was clearly stung by the experience.

If the point of the lawsuit was to stop her from expressing strong opinions, however, it had failed. She was a forceful, articulate voice against animal cruelty, wherever she encountered it. "When I started to get into the dark side of the zoo and aquarium industry," she told me, "particularly the marine mammal captivity industry—it started with Taiji, understanding what went on there—I mean, it makes the drug or mafia underworld look like a picnic. These people are blood brokers!"

Marino was born and raised in Brooklyn, the oldest of two daughters in a traditional Italian family. She has a don't-mess-with-me New York accent and an expressive way of speaking. Her voice roams across octaves. From an early age, Marino knew that science was her path, though she didn't initially set out to study dolphins. The insects she found in her backyard, the family cat, her tank of guppies, the stars in the night sky—every creature, every question, every last bit of the natural world, *everything* enthralled her. Marino's was a childhood filled with home telescopes, behavioral experiments with earthworms, wildly competitive science fairs: "the whole geek thing." Did life forms exist on other planets? If yes, how would we talk to them? Did dogs dream? What was the mean size of a millipede? What was it like to be a bee? Her inquiries began.

As a graduate student, Marino first glimpsed the dolphin brain in books; at the Smithsonian Institution, while collecting data for her PhD dissertation, she encountered actual specimens. More than anything, Marino was struck by the brain's uniqueness—it was oversized, rounder . . . *different*. The dolphins' ancestors had slipped into the water 55 million years ago, embarking on an evolutionary itinerary all their own. While humans took an express train and shot to their destination in no time flat, dolphins wound their way through geologic time, stopping often to take in their surroundings. In the end we both arrived at the same place—remarkable intelligence—but carrying different luggage. While everyone else in her field gravitated toward chimpanzees and other apes, the animals most reminiscent of ourselves, Marino was drawn to the weird, unfamiliar, and far more ancient architecture of the dolphin brain. "We're primates—I get it," she said, with a shrug. "But there is more than one way to be smart."

Marino was aware, of course, of John Lilly's similar pronouncement, his celebrated swerve from human to dolphin neuroscience decades earlier, but in her case he was no inspiration. "When I was starting out, people used to bring up John Lilly," she said, in an exasperated tone. "I had to fight for credibility to not be seen as a person who was just taking mushrooms and thinking dolphins were angels and all this crap. He really did a number on the profession."

Marino and her colleagues didn't have their own Caribbean research labs, but they did have CT scans and MRI scans and other technologies that helped them delve into formerly unknown territories of the dolphin brain. In Lilly's era, dissection was the main tool for investigating neuroanatomy, but that method had its limitations. It was like trying to wrangle Jell-O. (Large brains, in particular, are fragile and hard to keep intact.) Imaging solved that problem, allowing scientists to examine and rotate and peer into the entire structure in minute detail—charting the subtle diversities between sections, the exact measurements of regions, the precise stratum of cells. As our atlas of the dolphin brain is written, its terra incognita slowly becoming known to us, we are discovering that it is every bit as extraordinary as our own. "There's folklore about dolphin brains that says they're big but kind of simple," Marino said, shaking her head. "That's old stuff. We know now that it is a very complicated brain with a very wide range of types of cells. Their wiring's different. But it is just as complex—it could be *more* complex."

One of the most striking things about the dolphin brain is that its neocortex—the most recently evolved part of the mammalian brain that enables us to do sophisticated stuff like reason, use our senses, consciously think, socialize—is constructed in an utterly original way. It's a formidable structure: in humans, this area occupies 80 percent of our brain's volume. "There's a basic plan for the neocortex," Marino explained. "In all mammals it's layered." Ours is made up of six distinct layers, each of which contain specific cells that interpret certain types of information. But dolphins and whales have only five layers. "They're missing Layer 4," she said. "And the reason that's such a big deal is because, in primates, Layer 4 is where all the input for the lower parts of the brain come into the neocortex and get integrated." She raised her eyebrows. "So if they don't have a Layer 4, where's the information coming in?" There were some theories, she added, but no one really knew the answer. "The way information enters their brain, gets tossed around, and out of their brain—it's *completely* different." Her voice trailed off in a theatrical whisper.

If you pulled a neocortex out of a human or dolphin head, Marino told me, you could unfold it like a sheet. Ours is thicker, but theirs covers more real estate. The dolphin neocortex has more peaks and valleys, more crimps and wrinkles, more surface area for action to happen. In their brains, the region that deals with hearing is located at the top of their heads, while our hearing is processed in the temporal lobe, at the side of our heads. Dolphins have also rearranged the way they integrate sound and visual input. Instead of having that information zinging between the temporal and occipital lobes as it's being analyzed, going for a bit of a ride the way it does in our brains, their processing areas are located right next to each other, resulting in lightning-quick responses. If you were designing a high-performance computer, you would choose the dolphins' schematic, hands down. "This is a brain that is built for speed," Marino said, admiringly. "The rate at which they process information is *astounding*. Everything's faster! Their auditory fiber track is the diameter of this table!" This was said as a joke, a colorful exaggeration, but Marino was making an important point: "The bigger the fibers, the faster they conduct." She leaned back and smiled, her eyes wide with awe. "I mean, *are you kidding me?* We can't even imagine."

A fabulous neocortex is a kind of killer app for brainy animals, enabling the refined thinking and behavior that characterizes us as humans and distinguishes us from, say, lizards. It's where we get our abilities to make tools, use language, devise plans. When Marino talks about the dolphin neocortex, enthusiasm pours out of her: "It's got all kinds of different cells. There are columns, there are modules, shapes, clusters of cells. The *structure* in it! So many goodies. So much good stuff."

One group of brain cells neuroscientists find especially intriguing are spindle cells, also known as von Economo neurons (VENs). Humans and dolphins both have these cells in the areas responsible for high-level functions like judgment, intuition, and awareness—and so do whales, elephants, great apes, and even, it was recently discovered, macaque monkeys—but in the animal kingdom, VENs are unusual. Only the creatures with the most elaborate brains come equipped with them. Even their appearance is exotic: while many neurons look like wonky starbursts, their dendrite arms reaching across synapses to send and receive signals from nearby cells, VENs shoot out like bolts of forked lightning. They are also about four times bigger than most other brain cells. "They're like superstar neurons," Marino said. "And we see them in very interesting parts of the brain."

In new studies, researchers learned that when enough VENs are damaged in a human's brain, dementia can result. Losing even a portion of these cells causes us to implode socially, to lose touch with all niceties. We seem to need our VENs to get along with one another, to empathize, to know if we've made a mistake, to modulate our emotions. VENs play a role, it appears, in our ability to trust, to joke around, even to love one another. Whatever their purpose, early estimates suggest that dolphins and whales have about three times more of these superstar neurons than we do. Marino and other researchers suspect VENs are an adaptation that arose in big-brained animals to help shunt large parcels of information around at high speeds: "It seems to be something that emerges when you go from a certain brain size up."

One of the key reasons anyone's brain ballooned in the first place, scientists believe, was to deal with the intricacies of a thriving social life. Keeping track of family and friends and acquaintances in an extended community, figuring out who owes whom a favor and who once betrayed the group and who treated your grandma with special kindness but is also related to the guy who stole your brother's girlfriend—the fine web of interactions between hundreds of individuals—is as challenging for dolphins as it is for us. We need every bit of our brainpower to navigate these relationships, using everything from memory to judgment to communication skills (even *with* Facebook). Dolphins juggle not only close alliances within their groups, but also form alliances with other alliances.

So is their extroverted nature the reason dolphins developed such big brains? Likely, Marino said, but it's not quite that simple. "When you say 'social,' well, you have to have good communication to have a social culture," she said. "You have to have this, you have to have that. So it all gets wound together. And we'll never know because you can't directly test evolutionary hypotheses about complex behavior or cognition. But it's probably the best story we have right now." Doesn't it make sense, I asked, that when dolphins became physically less ferocious, when their bodies became smaller and their teeth less intimidating, that they would begin to rely more on the group? "They would need each other," Marino agreed. "Yeah."

In fact, dolphins are so tightly bound to their pods that they may be operating with a degree of interconnectedness far deeper than our own. "When you look at their brain you can definitely see how this could be an animal that takes sociality to another level," Marino said, pointing out that scientists can't explain why dolphins and whales strand en masse when only one or two individuals are sick; or why, when they're herded into the cove, they huddle together and don't jump the nets. "There is some sort of cohesiveness in them that I don't think we get quite yet, but it accounts for a lot of the behavior that seems strange to us." She took a sip of her tea and leaned over to pet a drooling St. Bernard that had wandered up to our table. In a field behind us, two horses cantered across the grass, whinnying and shaking their manes. "I think a lot of it comes down to emotional attachment," she continued. "And I think there is a very strong sense in them that if something happens to the group, it happens to *you*." She paused, and chose her words carefully: "I think the differentiation isn't that great between self and other."

≈

The possibility of a dolphin collective soul (my words, not Marino's) is an astonishing idea, but not a brand-new one. It was first proposed in the eighties by paleoneurologist Harry Jerison, who studied brain evolution and its effects on intelligence and consciousness. Jerison, who was obviously unafraid to tackle life's thorniest philosophical questions as part of his research, referred to it as "the communal self."

In this model an individual dolphin isn't so rigidly defined; he doesn't necessarily stop at the perimeter of his own body. His awareness, his concerns, even his survival instincts extend out into the world around him. He would relate to others in his pod at a level beyond empathy, in a kind of shared existence that we can't fathom.

The dolphin's limbic system, Marino told me, might well have adapted for this type of connectivity. It is an ancient part of the brain: the seat of emotions, memory, and smell. While most vertebrates evolved this region early and kept it pretty much intact, once again the dolphins came up with their own design. Since odors are indistinguishable underwater, their hippocampus, a region linked to their olfactory sense, diminished. Meanwhile, their paralimbic area grew huge, so densely jammed with neurons that it blurped out an extra lobe. There's a jubilee of tissue packed into this area, an exuberance of gray matter that scientists believe relates to all things *feeling*—and no other mammal has anything quite like it. In parts of the dolphins' limbic system the structure erupts in whorls and curlicues, like baroque décor picked out by Marie Antoinette. "It suggests that these animals are doing something very sophisticated or complex while they're processing emotions," Marino said.

That's the thing about brains. You can guess what they're up to, but right now, anyway, we really don't know for sure. The human brain contains hundreds of billions of cells busily engaged in thousands of trillions of unknown, vital tasks. Only the universe itself can rival its incomprehensible dimensions. We stand about as much chance of deciphering our brain's every last secret as we do of sitting down for a Starbucks venti latte with God. "There is no neuroscientist on the planet that can claim he or she knows how we go from this gray matter to being conscious," Marino said. "*Nobody* knows. It is a complete mystery."

Yet we're not entirely clueless. The anatomy itself has yielded rich information, tantalizing hints, and magnificent conjecture. We can examine how dolphin or human brains are constructed, and then match that to behavior. We can infer that a large brain size relative to body size makes for a more intelligent creature, although what *intelligent* means, exactly, is notoriously hard to define. It's undeniable that pea-brained critters like crows or octopi are capable of some awfully clever feats, while humans, with our cantaloupe-size heads, engage in all kinds of self-destructive nonsense. "The human brain is the most unsuccessful adaptation ever to appear in the history of life on earth," whale scientist Roger Payne once suggested. "What we call intelligence may only be a form of vandalism, just mischief on a grand scale." Trying to rank dolphin intelligence against human intelligence is like comparing submarines to airplanes, or the color pink to the color purple. They can't write things down; we suck at sonar. Rating animals' brainpower is a slippery task.

That doesn't matter, Marino said. We need to try. "Everybody knows intelligence is a fuzzy concept. What we have to do is make sure that we're being species appropriate and entirely empirical. In other words: dolphins recognize themselves in mirrors. Does this mean they are more brilliant than dogs, who do not? I don't know. I just know this is a capacity they have that many other animals don't have. And it means something."

The day had grown hotter as we talked, so we moved to a shaded patio overlooking a five-star canyon. A vegetarian lunch buffet had been laid out and was attracting numerous takers: high-school volunteers wearing Best Friends' T-shirts, staff members emerging from offices trailing beagles and collies. Marino and I stopped to get salads. "The person who has probably done the most to help us understand dolphin intelligence is Lou Herman," she said, after we sat down with our plates. "His work was *superb*."

Louis Herman is a psychology professor emeritus at the University of Hawaii. His studies on dolphin cognition, perception, memory, and communication were groundbreaking, jaw-dropping—any awesome adjective you'd care to insert. With scientific rigor and a lot of creativity, Herman showed just how smart dolphins can be. "My thought was, 'Okay, so you have this pretty brain. Let's see what you can do with it,'" he told *National Geographic*.

Herman's research, at the Kewalo Basin Marine Mammal Laboratory, was based in Honolulu. From 1970 to 2004, he worked with bottlenose dolphins, teaching them a gestural sign language and another language based on sounds, and then testing how well they grasped various concepts—including many tricky abstract notions that animals are really not supposed to be able to understand.

Apparently, no one had informed Herman's dolphins of this. His bottlenoses responded to complicated sentences and knew exactly how the word order, or syntax, changed their meanings. They got instantly that a command like "take the surfboard to the Frisbee" was different than "take the Frisbee to the surfboard," and they adjusted their movements accordingly. "An important finding," Herman wrote, "revealing the dolphin's mastery of the sentence forms used, was that understanding was shown for novel instructions as well as for more familiar ones, with only a slight advantage to the latter." When the dolphins were asked to do something impossible, like bring the tank window to the surfboard, they wouldn't attempt any action. They just stared at their trainers, as if to say, "*Come on*. You and I both know that can't be done."

Herman's dolphins could distinguish left from right, even when the directions were reversed arbitrarily. They understood the concepts of presence and absence. Using their flippers to press paddles that indicated yes or no, the dolphins correctly responded when asked if a person or object—a boy, a particular dolphin, a ball, a box—was in their tank or not. They could listen to a series of eight sounds and then indicate whether a ninth sound had been previously played. They discerned the ideas of "same" or "different," and "less" or "more." In other studies, they reported whether they were "sure" or "unsure" of an answer to a difficult question. When asked to create new behaviors—tricks they'd never done before—they immediately began to innovate, in perfect unison.

Herman also proved that dolphins understand what we mean when we're pointing; that they can identify their own body parts; that they realize television is a representation of reality. They can remember objects, locations, and instructions even over time, recalling information as necessary. They are masterful mimics, able to imitate sounds and movements with ease. With no prior instruction, when a trainer lifted a leg in the air and asked the dolphin to do the same, the dolphin lifted her tail. None of this is simple or easy behavior; all of it demonstrates, as Herman concluded, "wide-ranging intellectual competencies . . . an intellect that meets with some of the hallmarks of human intelligence."

Not only were the dolphins capable of all this (and probably far more), they picked everything up at warp speed. "When you work with them, you realize they're on a completely different temporal plane than you," Marino said. "They're always one step ahead. They get things much quicker. They do things much faster." She laughed. "I mean, you can tell they're impatient because they have to deal with slow humans."

≈

The night before I flew out I stayed in Las Vegas, doing some lackluster gambling and wishing I had a communal self so I could tell how the blackjack dealer was feeling about his hand. After an improbable win at slots, I quit while ahead and went back to my room. My own brain was spinning after talking to Marino. I felt as though I could have stayed at Best Friends for months, volunteering to clean out the pig pens if it meant the chance to ask Marino more questions about the inner lives of the sanctuary's animals.

Later, Marino would e-mail me in response to queries, and I took every exchange as an opportunity to probe further into her thoughts about other creatures, how their brains might inform their worlds. My longing to know more about animal minds was far from academic. Recently, my two cats, Mouse and Georgia, had both died suddenly of a virulent disease. Losing them felt like life's sucker punch, and returning home to my apartment after they were gone, I was devastated by its emptiness. Attachments with animals run deeply through our bedrock, and we have no reason to think that creatures with different brains than our own feel any less connected than we do. This is especially true of dolphins. In fact, their elaborate limbic system might make them feel the loss of another even more acutely.

On my last night in Vegas, sleep came quickly but strangely. In my dreams I found myself adrift in a stormy sea at twilight, wind howling and waves battering rocks, while all around me the black dorsal fins of dolphins rose and fell in the tumult, coming precariously close to shore. On a cliff above, shadowy figures lurked: it was important to escape them. To avoid the pounding surf I dove, and for some reason I could see perfectly underwater, as though I were wearing goggles. Below the surface, the dolphins were trying to get my attention. They were excited, gathered around a doorway under the ocean. The door was bright turquoise and quite small, as though built for hobbits (or dolphins). They pushed their beaks against it and nudged me forward. I opened the door and swam through. On the other side, the water was darker, the color draining from marine blue to inky navy to a blackness that felt absolutely still. Then I realized I wasn't in water any longer, but in space. The medium was rich, as viscous as oil, suffused with grandeur and an almost unbearable sorrow. I swam deeper and deeper, beyond anything I could recognize but the fast sweep of dolphins moving around me in the beautiful, terrible void, my own heartbeat, and the winking of the luminescent stars.

CHAPTER 7

HIGH FREQUENCY

I was in Los Angeles when Joan Ocean e-mailed, inviting me back to Hawaii. If I could come to Kona the following week, she wrote, I could be her guest at a sold-out gathering she was hosting, a five-day workshop called "Dolphins, Teleportation, and Time Travel." During the day, we would swim with wild dolphins along the coast. In the evenings, we would listen to lectures about dolphins. Throughout the week, the group would remain in "dolphin consciousness," which Ocean described as an "active, relaxed, hyper-aware state of high intelligence." There would be guided meditations, fire circles. "The dolphins have been amazing lately," she added. "I don't like to always use that word, but it definitely applies here."

I packed enthusiastically. I was still curious about why dolphins were such key players in the New Age movement; why the words "teleportation," "time travel," and "dolphins" might appear in the same sentence. When I'd been to Dolphinville previously, I had only dipped into Ocean's world. This would be more like a dunking, in which I'd be surrounded by people with extraordinary—some would say zany—beliefs. In their lexicon, dolphins are not mere animals. They are beings from another dimension, visitors from faraway stars, wise elders here to teach us vital lessons. At Ocean's seminar, my fellow attendees would be people who see these animals not as a scientific challenge—where *is* that Level 4 in their neocortex?—but as one great big existential riddle.

Can dolphins swim through space and time? Can they show us how to love? Do they know something about life that we do not? (The dolphins in Douglas Adams's novel *So Long and Thanks for All the Fish* certainly did, zipping off the planet just before an imminent doomsday, leaving behind only a fruit basket and a thank-you note.) As far-fetched as these questions may seem, ever since humans met dolphins we have felt a compelling attraction to them, a kinship we can't entirely explain. This is the connection John Lilly tapped into; it was the source of his immense public popularity and also what spun him out, and whatever you want to call it, this inkling or feeling or hope that dolphins are some kind of aquatic oracles is widespread. It's common. To scientists' dismay, it verges on conventional wisdom. What Ocean was convening was the opposite of an academic gathering. It was a fantasy camp for imaginative dolphin lovers, a sort of Dolphin Comic Con. Which, to me, seemed like a wonderful reason to go.

My plan was to keep an open mind. I would be an amiable but neutral observer. To a point, anyway: convincing me that dolphins come from another solar system, or that anyone, to date, has time traveled, would be an uphill battle. But already, to a less astral degree, I had been forming my own strong feelings about dolphins. There *is* something singular about them, and I've felt it every time I've been in their presence. I have heard this chalked up to recognition—that we detect the same spark of higher intelligence in them that we find in ourselves—

but orangutans are wicked smart too, and you don't find people gathering to teleport with them.

Dolphins are enigmas. Maybe we're hoping they *do* have helpful intergalactic know-how; some secrets to living in harmony would be timely right about now. Maybe we have no idea who dolphins are or what they're up to or what they're capable of, so we make up stories to soothe ourselves; maybe our own brains need to bake for another fifty or sixty million years before we can relate to them fully. Or maybe, as some researchers assert, dolphins are nothing special. Maybe they're overrated and only about as clever as parrots, and in a well-intentioned but misguided way, we project our spiritual longing onto them. Maybe we are hardwired to love any animal that looks like it's always smiling.

The dolphin-as-dullard idea has been vigorously refuted by all but a few scientists, but I include it as a reminder that we don't have everything figured out. We reckoned only four centuries ago that the Earth revolved around the sun; as recently as 1850, no one believed that germs cause disease. The cycle goes like this: we're sure of something—and then we're not. We correct ourselves and carry on. If you could plot on a graph our attempts to catalog our world, it would be a series of short, sharp zigzags running all over the place. "Earth, to put the matter succinctly, is a little known planet," E. O. Wilson wrote. The question I was turning over and over in my mind was this: What *don't* we know about dolphins? For that matter, what don't we know about ourselves? What don't we know about everything? The answer to all of the above is: plenty.

You might be pleased to learn, for instance, that rather than the five senses you think you possess—plus a "sixth sense" if you count intuition—humans have at least twenty-one means of perception. Our biological toolkit includes proprioception (the position of one's body in space), chronoception (a sense of the passage of time), nociception (the awareness of pain), equilibrioception (if you've ever had vertigo, you know what it means to lose this), and thermoception (a sense of hot and cold), among others. There are internal sensors throughout our bodies—in our brains, hearts, blood, skin, cells—registering even the most ethereal cues. One recently discovered sense is magnetoreception, or the ability to track the planet's magnetic fields. We may have this sense, weakly. Or we may not. Dolphins definitely have it. Sharks, birds, sea turtles, bats, butterflies, and honeybees, to name just a few animals, use it to navigate. No one knows exactly how it works, but scientists think that magnetite crystals in the creatures' heads might subtly pull them in one direction or another, guiding them across great distances with surreal precision.

In other superpower news, people and monkeys have managed to move objects using only their thoughts. Dolphins communicate through their foreheads. Birds can feel earthquakes coming hours in advance. Prayer has healed people. Brain surgery patients placed into temporary comas were able to recall—in astonishing detail—conversations that occurred in the operating room while they were technically dead. Neurologist Allan J. Hamilton, quoted on Harvard Medical School's Web site, recounted one of these incidents, asking: "What, I wondered, should those of us in the medical field do with such unsettling disturbances, such seeming ripples of the supernatural? Ignore them? Or should we declare them simply to be a puzzling mixture of science and spirit? Can we now allow ourselves to entertain the possibility that the supernatural, the divine, and the magical may all underlie our physical world?" So what other exquisite senses and abilities might humans—or dolphins—have that we don't even realize? The zigzags continue.

"Not only is the universe stranger than we imagine, it is stranger than we can imagine," the British astrophysicist Arthur Eddington famously declared. I knew those would be worthwhile words to keep in mind while attending sessions with titles like "The Parallel Worlds of Our Multiverse." "Dolphins play with the energies they attract," Ocean writes in her book *Dolphins into the Future*. "Maybe they are playing with past, future and parallel lives, with fourth-dimensional ghosts, the energies of the cetaceans in the Arctic, the inner earth, human auric fields, colorful rainbows, or music or entities living on other planets. They can access it all. What a fun way to live! No wonder they seem so joyful."

It's easy to dismiss what you don't understand. It's unnerving to think that the world might be weirder than advertised. Daily, we forget how miraculous everything is, down to the teeniest subatomic speck. According to quantum theory, the universe is a seething field of potential, constantly creating itself—and what we consider reality is only a slightly more convincing type of dream. If you swap-in a different brain, your reality changes too. And these are findings that we *have* proven. Why should anything surprise us?

≈

"I know many of you have been here before and have heard the basic dolphin information," Ocean said, surveying the packed room, people staked out in chairs, on the floor, anywhere they could find a seat. "So tonight I thought I'd talk about the far-out stuff." She beamed; the crowd whooped. "We can take it, Joan!" a man wearing a Sasquatch T-shirt shouted. "We're ready!" a woman with a magenta streak in her hair yelled. A crystal dolphin sculpture, leaping from its pedestal, cast light patterns on the wall behind her.

It was Day 3 of the workshop and I had just arrived, pulling into Sky Island Ranch as the evening's program began. I wedged myself into the room and looked around. There were at least seventy people here, slightly more women than men, a wide span of ages and nationalities. There were Australians, Brits, and Germans. A South African and a New Zealander. Several Canadians. A handful of kids under ten. Dolphinville was well represented: I recognized a local divemaster, an underwater photographer, and a boat captain. It was an attractive crowd, not a scruffy hippy or burnt-out necromancer in the house. They'd spent the morning on the water, Ocean said, and swum non-stop with the spinners.

Ocean leaned back in her chair, a white swivel lounger. Earlier, she had told me that she calibrated her talks according to who was in the audience, "to know how much to share." In this room, there was no need to hold anything back. She could lob out phrases like "holographic communication" and "fourth-dimensional beings" and know that nobody would scoff. "There is a longing in people to have contact with dolphins," she began. "It's not some whim. We have a deep soul connection."

Her own dolphin experiences began in 1978, at a workshop with John Lilly about out-of-body experiences. "I liked his sense of humor," she said. "Oh, boy. You just never knew what he was going to do next." Lilly played tapes of dolphin vocalizations for the group, day and night, at top volume. "So maybe that's where everything started," Ocean mused. "Maybe they were saying things and I was picking it up." Not long after that, she started receiving messages from the dolphins themselves. For the past thirty-five years, she said, she'd been in close communication with them, swimming with twenty-eight species of dolphins and whales in twenty countries. The spinners in Hawaii were the ones who had divulged the most. "We are here to teach you to move beyond the limits of the five senses," Ocean felt the dolphins tell her as she swam in Kealake'akua Bay. "We encourage you to communicate with us in the unexplored domains of the sixth sense and beyond." Plus, the dolphins added, it was possible to merge the senses together: "You will begin to smell images, hear feelings, and see sounds."

Though such a mash-up might sound uncanny, it actually exists, and is referred to by scientists as "synesthesia." People with this neurological condition—and there are a fair number of them—experience such bizarrities as feeling colors, tasting shapes, scenting emotions, and viewing numbers as structures. It's even more common if you canvas the creative fields: the artist David Hockney has synesthesia, and so did writers Charles Baudelaire, Arthur Rimbaud, and Vladimir Nabokov. The inventor Nikola Tesla was said to possess it; musicians as diverse as Franz Liszt and Pharrell Williams have claimed it too. Research has suggested that we all have this vibrant cross-wiring at birth, but for reasons unknown, most of us shed it fairly quickly. Others maintain it throughout their lives, and dwell in a far more textured world; my readings about the syndrome made me wish that I had it.

As the room digested Ocean's revelations, she continued. "I was swimming with the dolphins for about two months before they included me in their pod," she said. "The pod field of energy can be very wide. Remember, they can communicate over long distances to each other, so they don't need to be side-by-side like we do." As Ocean glided among them in a relaxed state, she would ask questions and then wait for images to pop into her mind, which she took as answers. Among the unexpected things the spinners told her was that she should pay more attention to opera.

Ocean bought recordings of the great sopranos and listened to them constantly, trying to hit some of the high notes herself. Compared to dolphins, however, we are all baritones. They hear sounds up to 160 kHz, eight times higher than we do. Dolphins' ultrasonic capabilities, Ocean believes, are not just a means for the animals to navigate and hunt fish. She describes their sonar as an advanced form of expression that can alter reality, opening up portals into other dimensions. "These tones can transform all things," she has written. "They can heal and change our bodies and our environments. They can dematerialize and materialize matter, and even change the physical structure of objects (demonstrated in the third dimension by sound that can shatter glass)."

The New Age world is full of references to frequency and vibration, the idea being that higher vibrations represent love and transcendence, and lower vibrations are the dull and sluggish stuff of negativity and disease. Even if you cringe from talk of "chakra balancing" or "harmonic healing," you're probably aware that everything around us is oscillating at all times, even the heaviest solid objects, and that energy waves can have exceptional force, even if we can't see them or hear them (think earthquake, laser, microwave oven). Sound as treatment is a fast-growing field: at frequencies far above what dolphins can generate, ultrasound has destroyed tumor cells, healed broken bones, and cauterized wounds. It can wipe out toxic algal blooms in lakes and oceans, and cause drops of alcohol, plastic beads, matchsticks—and in one boisterous experiment, the ingredients in a caprese salad—to levitate. Blasts of high-frequency sound waves *can*, as Ocean claims, change the physical property of a substance. Liquid can turn to jelly. Bacteria can disintegrate. Water can be zapped into mist.

Even at lower frequencies, sound has intense effects. A major study at London's Chelsea and Westminster Hospital found that when lovely music was piped into their rooms, patients healed faster. Also, the staff was happier, neonatal infants thrived, everyone's blood pressure was lowered, and surgeons performed more effectively. When the results of this experiment came in, the hospital recruited Brian Eno, a musician who has worked with David Bowie, U2, and Coldplay—to help compose its emergency room's playlists. On the flip side, the imaginative folks at Raytheon devised an acoustic riot shield that uses low-frequency sound waves to interfere with the respiratory tract, incapacitating protesters by making it hard for them to breathe.

None of this proves, of course, that dolphins are using their echolocation to "stop time" or "transport themselves into other realities," as Ocean believes; if they are, then bats must be doing it too, because their frequency range is even higher, up to 212 kHz. But sound is powerful, and to imagine that dolphins, maestros of underwater acoustics, might have some abilities in this realm that we haven't guessed is not an unreasonable stretch.

One side effect of the notion that dolphins' sonar has transformative powers is that a shady industry promising "dolphin therapy" has sprung up to capitalize on it, charging astronomical fees. Marketed to parents of kids with autism, cerebral palsy, and a host of other afflictions ranging from quadriplegia to bed-wetting, it costs about $3,500 per week. I'd talked to Lori Marino about dolphin therapy when I was in Utah; she had coauthored a paper that investigated its healing claims and showed them to be scientifically unsupported. Any benefits are short-lived, she found, fading fast when the family returns home—adopting a puppy would have more enduring results. "It's the worst type of snake oil you can imagine," Marino said, "taking advantage of desperate people who have a sick child and telling them, 'Yeah, if you give me enough money and the dolphins swim with your kid for a half hour every day, that child is going to be helped.' It's a global sham."

Ocean had another perspective. "It's not that dolphins heal people," she told the group. "It's that being with them helps people regain their natural healthy state." They lead us beyond our limitations, she explained, into "a universal field of energy that is the source of all possibilities." To her mind, they don't fix us so much as take us into a place where we don't need fixing. One time, she recalled, a dolphin swam up to her and loudly sonared her ear, which happened to be infected. "I wasn't even paying attention," she said. "He came right at me and . . ." She imitated a high-pitched dolphin whistle: "Eeeeeeeeeeeeeeeek!" The dolphin's treatment worked, she said: "I felt the ear open up." She urged us to ask the spinners for help if we needed it. But the request had to be made from the heart, she stressed. "What carries the communication are your feelings of love. Feelings of caring. And you know that feeling when it comes over you. You think of the dolphins and you just feel so . . . much . . . love." Her voice cracked with emotion. "When you're feeling that love in your heart—*that's* when you're communicating."

I glanced at everyone. Several people had tears in their eyes; the children were sitting absolutely still, staring at Ocean with rapt attention. There was a vibe in the room that was sort of intoxicating. I was startled to recognize the feeling: it was the same Zen calm that came over me when I was in the ocean with the spinners. It was the bone-deep peace that arises during meditation, the incandescent warmth you feel around someone you love. It was the opposite of aggravation. My usual life sound track, a low hum of anxiety, had simply stopped. Was this dolphin consciousness? If so, we should try it more often.

Ocean continued to talk about her experiences with the dolphins; her conviction that they were not just out there chasing fish, but actively transmitting knowledge. "When they are playing with us," she said, "they are giving us information." A man with a goatee piped up, asking for specifics. "There are some far-out things," Ocean said, "but . . . those are too far out." "Don't stop now!" someone pleaded. "Well," Ocean said, hesitating only for a moment. "There was one thing I wouldn't put in my book. I was afraid I'd lose all credibility." The room hooted encouragement. "They said there are ET vehicles underwater," she continued. "They travel the waters of the planet and there are docking places where the dolphins can actually swim inside these spaceships. They showed me a picture of them."

"Are the ships over at that bay we call the End of the World?" a ponytailed woman wearing a tie-dye sweater asked. "I heard that's a place where they go."

"They're pretty much wherever they want to be," Ocean said. "I mean they probably need some amount of depth but . . . have you seen them there?"

A number of hands shot up. "I know people who have seen them come out of the water," a man with a German accent said, his pupils dilated with excitement.

"We've seen what I would call plasma ships," Ocean confirmed. She pointed to a group of Dolphinville residents who were nodding vehemently. "We were out on the boat together and we all saw it, this giant white pulsing form that was about the size of a football field. We could swim into it—if you were in it you could see yourself, but if you stuck part of yourself out you couldn't see that part. It was the most ecstatic feeling."

"The dolphins were swimming through it," Ocean's friend Celeste added. Celeste was in her seventies, with long white hair and an elegant manner. She was the most athletic-looking septuagenarian I'd ever seen, besides Ocean. "But these dolphins—they weren't *our* dolphins. They looked like spinners but then they didn't once you got into the plasma. They didn't have a mark on them, and they were lighter. They weren't from here."

"*Everything* was different about them," another woman agreed emphatically. "They were not our dolphins."

The room erupted in murmurs. "Wowwwww," said a young woman with a flock of birds tattooed across her shoulders. "Someone in Hilo told me he met a mermaid," another guy offered. He was dark-eyed and rangy, folded into the corner but eager to share. "She was ten feet tall and she came out of the water and she had big teeth and long hair."

Ocean nodded. "Interesting."

"She said she lives off Maui," the guy added.

"I think they're going to make themselves known, more and more," Ocean said. "I think they're being brought in now because we're open to them. So there will be sightings. It's great."

"All the channels are open," the guy agreed.

The discussion of underwater UFOs had turned the conversation away from dolphins, but Ocean steered it back. We would wrap up with a "jump" into their realm, she said, a meditation in preparation for meeting them tomorrow. As everyone shut their eyes and reclined against cushions, she turned on some seriously trippy music, the notes soaring and diving. I felt woozy, as though my head were melting. "Okayyyy," Ocean began, in a purring voice. "Please relax as we prepare to enter another timeline. For now, be aware of each other. Kindred souls, joining together with pure intentions. Expressing our love and peacefulness. As you breathe deeply, feel the love enlightening all of your body. Your bones, your muscles, all of your organs. Your bloodstream and your nervous system. All activated into pure love . . ." The music swooned. Some people had slid down so they were laid out on the floor. A ceiling fan ticked overhead.

Somewhere in the near distance, I heard chickens. The only other noise was the shushing of palm trees tossing in the wind. Ocean spoke in low, hypnotic tones: "Now, we send our love into the ocean. We ask the dolphins to be with us, to play with us in the beautiful waters here. We connect with the consciousness of Mother Ocean. The dolphins are sending their frequencies back to us, and in this moment we give thanks to them for all the joy that they bring, for their example of kindness, compassion, family living, connection to their environment. We are grateful. And we look forward to swimming together with them— our pod and their pods, together."

I felt myself smile. Spaceships or not, it was hard to argue with that.

≈≈

Just after dawn the next morning I drove to Honokohau Harbor, watching the sky turn from navy pink to apricot blue to lavender gold. At the docks, we split into two groups. Most people boarded a double-decker boat that could comfortably hold everyone, but Ocean waved me onto a smaller vessel, a powerboat that only took eight. On deck I could see Jan, a lithe diver and skipper, casting off the ropes, and Celeste, her silver hair in a sporty braid, storing her gear like a pro. It was Day 4 and Ocean, I could tell, was angling for a respite, a morning of restoration among the spinners, without having to answer dolphin or UFO questions every ten seconds. While the other boat loaded its heap of snorkeling gear, our captain, an easygoing, ample-bellied man named Kit, suggested that we shove off immediately, because fishing boats had radioed a sighting of a pod of pilot whales cruising five miles offshore.

Without much notice to the others, we headed out. Searching for a pod of pilot whales on the move is a long-shot proposition, but Kit ventured west with purpose. Somehow, I knew we would find them. I was thinking about what else we might encounter this far out, and watching the profound blue of the deep ocean overtake us as we sped away from land, when the radio squawked again: another sighting, closer. We drove about a mile in that direction and stopped, rocking on a light swell. Everything was quiet, the ocean luffing against the hull.

"Whoooooofff!" A pilot whale surfaced with a sudden gust of air, a hundred yards away from us. "Poooooooshhh!" Another whale came up right beside him. Their black backs rolled and dived; we could see their bulbous heads, their swept-back dorsal fins and missile-thick bodies. Then, more fins, all around. These animals were three times the size of the average spinner. "There are at least forty in the pod," Ocean counted, doing a 360-degree scan. "And they have some calves with them."

Kit drove ahead; he would position us far in front of the pod so we could jump in and swim with them as they passed. I fiddled with my mask, feeling a steady adrenaline drip. I'd never been frightened around dolphins, but at twenty-feet long and tipping in at three tons, these were more like orcas. Also, I was aware of pilot whales' reputation as stubborn, rather sulky animals who've been known to express their displeasure in creative ways. I thought back to one unpleasant incident that occurred in these waters, under more or less these exact circumstances, when a pilot whale had snatched a swimmer by the ankle and yanked her forty feet down. The woman barely escaped drowning.

Still, I knew I couldn't miss the chance to observe them underwater. Compared to the spinners, the pilots had an air of gravitas. These were short-finned pilot whales, but there are also long-finned pilot whales, the two species closely resembling one another. Both were members—along with false killer whales, pygmy killer whales, melon-headed whales, and orcas—of the group of dolphins known as "blackfish." Like killer whales, pilot whale pods are focused around matriarchs, mammoth mommas who run the show, and whose sons stick with them for the long haul. Grandmothers, too, play a pivotal role: for up to fifteen years after their breeding years are over (at around age thirty-five), female pilot whales continue to lactate, and help nurse and baby-sit their podmates' calves, bolstering their survival chances.

Ocean, suited up and ready to launch off the stern, turned and in her husky, gentle voice informed me that pilot whales are usually accompanied by oceanic white-tip sharks, one of the nippier models. No one knows precisely why, but the two species often travel together. "So just keep an eye out behind you," she added, with a grin.

Kit cut the engines and told us we could slip in, two by two. The ocean was crystalline, a six-dimensional heaven of azure, lapis, and sapphire. Even with such lucid visibility there was no spying the seafloor out here: we floated in water miles deep. "Starboard!" Kit shouted, pointing. I saw a pair of whales surface nearby; they were headed straight for us. I dove and so did the pilots, and soon they loomed into view beneath me, two huge adults with a calf tucked below them, swimming by slowly and sounding me with their echolocation clicks and creaks and squeals, like an undersea radio receiver tuning into the snorkeler station. There was a stateliness to the pilots' movements, and they emanated waves of that entrancing calm I had become so fond of. Sunbeams played through the water and danced across their bodies like spotlights.

The whales dissolved into the blue and we clambered back on the boat; Kit shuttled us ahead and then dropped us in their path again. If bothered, the pilots could have simply dived out of sight: they can go fifteen minutes between breaths. One of their hunting strategies for squid, their main prey, is to sprint-dive at 20 mph down to 3,000 feet, an effort so strenuous they've been nicknamed "the cheetahs of the deep sea." Pilots are even thought to compete with sperm whales for the ultimate prize: Architeuthis, the giant squid. In the Canary Islands, they've been spotted swimming along with four-foot-long tentacles trailing out of their mouths.

The pilots didn't shy away from us; they passed closer this time, allowing the calf to slide out beside them. The mini-whale was curious, eyeing us with interest. I hovered, lost in awe. Time went sideways. At one point I realized that I'd forgotten to breathe. My lungs contracted but I felt no panic, even though I was twenty feet down and wearing a weight belt. Was I narked (divers' slang for drunk on nitrogen, bent on bubbles)? How could I be? I didn't have tanks. But that's how I felt. If more people swam with wild cetaceans, I thought as I kicked to the surface, we wouldn't need drugs. We wouldn't need legalized weed. We wouldn't need Xanax or Prozac or Ecstasy.

I made my way back to the boat and we set off again, following the pod and then overtaking them. The other dive boat was alongside us now, the others clamoring to get in with the whales too. At the first opportunity, dozens of people jumped into the water. The process took a while and by the time everyone was situated, the pilots were gone. I swam around searching for them, but all I could see was a blue infinity, and someone on the large boat swishing a pole camera around to investigate the scene. "Shark!" I heard one of the divemasters call out, but I couldn't spot it. Later, we would watch video footage of an oceanic white-tip knifing around us, just outside our field of vision.

While we were in the water, Kit heard radio reports of a spinner convention just up the coast, a gathering of three to four hundred dolphins. We split from the others again, jetting north toward Kohala to see if we could find them. As we motored, I told Ocean how blissfully untethered I'd felt among the pilot whales, unconcerned with banal things like gravity and time and even air. She nodded, laughing. "My whole goal is to get people into that place," she said. "It's love and gratitude. And it means a lot."

The sun glinted down on the water and the water sparkled like sun. As we drove, Jan passed around fresh mango and pineapple spears. I sat next to Chrissi, a soft-spoken Canadian in her early twenties with freckles and a tomboyish quiff of red hair. She had traveled here from Nelson, a hip, outdoorsy town in British Columbia. Like all of Ocean's followers, Chrissi had a colorful set of personal beliefs. "I was very interested in Sirius," she said, explaining why she'd been drawn to Hawaii. "I read everything I could get my hands on. Then I found Joan's Web site and I started reading about dolphin consciousness and I thought, 'This is perfect. This is like a dream come true.'" Swimming with the spinners, Chrissi told me, had helped her cope with crushing grief. Not long ago, she had lost her father; and then her beloved dog, Topaz—"my baby, my best friend, my everything"—had died too. In the face of this double-header heartbreak, Chrissi's outlook was so sweet, so relentlessly sunny and hopeful, that I felt humbled by meeting her.

The spinners were easy to locate. There were platoons of them and even from a distance they were visible, shooting skyward one after another, whirling through the air like corkscrew rockets and careening down with such dramatic splashes that they might have been competing in a spinner X Games. I had never seen them so lively. When we drew closer to the pod, which stretched as far as we could see, the dolphins shot to the front of the boat, jockeying to ride our bow waves. They came speeding in from all directions, and when we slowed they slowed along with us, circling the boat with a puppyish glee. We drifted in aquamarine water above a coral reef teeming with iridescent fish, and if the dolphins were enticing us to join them, they couldn't have chosen a more alluring spot. For hours we swam there, and so did they.

For hours, we played. Nearby, I saw Chrissi underwater, suspended between two dolphins who were imitating her movements. If she dove, they dove. If she hung in the water, so did they. I remembered how my first spinner encounter had jolted me out of the depths of sorrow; I hoped Chrissi was feeling the same effect.

The spinners were also following Ava, a soulful seven-year-old girl who had come with her mother, Suchi, a friend of Ocean's. Earlier, Ocean had referred affectionately to Ava as "dolphin bait," noting that children seemed to draw the spinners' attention. Dolphins have also been said to gravitate toward pregnant women, whom they can examine internally with their sonar. These anecdotal tales would seem innocent enough, and they might even be true, but not long before I arrived in Hawaii a controversy had broken out over one Virginia couple's decision to come to Pahoa, a rural town on the flank of the Kilauea volcano, in the hopes of delivering their baby in the ocean, attended by dolphin midwives. Somehow, their plan had made the press, and provoked heated discussion. DOLPHIN-ASSISTED BIRTH—POSSIBLY THE WORST IDEA EVER, reported *Discover* magazine's science blog, pointing out that dolphins regularly ingest creatures the size of a newborn; that tiger sharks, too, might be interested in participating when the blood and afterbirth started swirling around. It was also pointed out that the "beach" near Pahoa was actually lava rock covered with spiny sea urchins.

In the end the event didn't occur, although several commenters endorsed it, accusing the naysayers of being predictably negative humans. One woman wrote that her dream was to deliver her baby "in the presence of a pack of wolves, to ensure that my child has an intimate knowledge of nature from the very beginning." Another angrily recalled how she had been shunned for deciding to give birth in a cave "where a mother brown bear currently lives with her cubs," but confirmed that it had been "an amazing process." The anti-dolphin-midwife crowd, she said, were assholes. "Yeah, maybe," someone responded, "but I'll be the asshole whose baby isn't eaten by fucking dolphins."

It is hard to be objective around dolphins; our emotions are entangled from the moment we see them. Regardless of how spiritually salving it feels to hang out with them, these are undeniably large, professional predators who were, at this moment, supposed to be resting up for a long night of hunting. Their survival depends on their ability to take down the wiliest prey, in rough and roiling conditions that could pit them, at any moment, against razor-toothed sharks. Into this equation comes a batch of humans wanting to . . . commune. Was it equally awesome for them to be around us? I hoped so—but I was far from certain.

Afternoon was coming on and a shiver of wind had kicked up, so I swam back to the boat. Ocean was on deck, wrapped in a flowy cover-up decorated with vivid Pucci swirls. I stripped off my wetsuit and rinsed under a warm hose; soon the others were climbing out too. Jan pulled up the ladder and Kit started the engines to return to the harbor. On the way back we ate cookies and talked about transcendence. "People are evolving," Ocean said, contentedly. "It's just explosive right now. We're learning so much, so fast." Her hair whipped around in front of her face; the clouds reflected in her polarized sunglasses. "We *are* multidimensional beings. But if you don't use it, you'll lose it."

≈

If the New Age world has adopted dolphins as its totem animal, anointing them finned avatars of higher consciousness, revered creatures whose presence speaks of deeper meaning, it should be acknowledged that other, older groups arrived at this conclusion first. At least fifteen thousand years ago California's earliest inhabitants, the Native Americans of the Chumash nation, referred to themselves as "the Dolphin People." Their history spelled it out concisely: This tribe was not merely friendly to dolphins; they considered the dolphins to be their direct relatives. The Pacific Ocean was not just the monumental vista they gazed at from their villages; it was their ancestral home. The Chumash were hunters and gatherers, renowned for their weaving and bead-making skills, their prowess with agile, redwood-planked canoes called *tomols,* their peaceful, resourceful ways—and they were also full-on marine mystics.

Millennia ago, more than twenty thousand Chumash people lived in fire-lit settlements along the central and southern California coast, from Point Conception to Point Dume, as well as on the nearby Channel Islands: Santa Cruz, Santa Rosa, Anacapa, and San Miguel. Then the Spanish arrived in the eighteenth century, bringing conquest and disease and missionary zeal: by 1900 only two hundred Chumash were left. But despite this tragedy, the tribe's past has not been extinguished. Its heritage remains, and so does its bond with dolphins.

At Malibu's northern edge, in fact, next door to a phalanx of multimillion-dollar houses, there is a replica of an authentic Chumash village. Its name is Wishtoyo (the Chumash word for rainbow), and its cofounder and present-day steward, a Chumash man named Mati Waiya, has been actively reintroducing his culture's traditions to a world sorely in need of a refresher course.

I drove down the dirt road to the village on a late summer afternoon, the dust from my tires billowing red and gold in the dusky light. If you weren't looking for Wishtoyo, you might never find it—it isn't a showy place. Tucked next to a gulch on a cliff above the Pacific, its buildings blend into the landscape so thoroughly it's as though they've sprung from the earth itself, and actually, they have. Wishtoyo's six dome-shaped dwellings are made of willow branches and woven tule reeds, their entrances draped with deerskin and framed by whalebones and antlers, adorned with stones and seashells. Their floors are soft sand.

As I pulled in to park, three huge German shepherds came loping toward me, followed closely by Waiya himself, a tall and strapping man in his fifties. His waist-length black hair was pulled back and secured by two long slivers of bone. A smaller bone bisected his nose, sitting atop a sleek mustache, beard, and goatee that almost looked as though they were tattooed onto his skin. Waiya wore strands of beaded neck-laces, a tribal print sarong, and a sleeveless shirt that revealed thickly muscled arms. He was barefoot. A raptor claw hung from his left ear. If I hadn't already talked to Waiya on the phone and heard his voice, a smooth California drawl, I would have expected him to start speaking Smuwic or Samala or another one of the Chumash languages.

Waiya greeted me, and we launched into the stories. "We're a mar-itime people," he told me, as we walked around the village. "The dol-phins are our relatives, our brothers and sisters. *A'lul'koy* is our blue dolphin. Malibu is *Humaliwu,* where the waves crash loudly. *Muwu* is the big ocean. These are all Chumash names that people don't even know." He stopped in front of one of the buildings, which was about the size and shape of a large igloo. "It's called an *ap*," Waiya said. "A family would sleep in here." We ducked inside, followed by Sumo, one of the German shepherds. The air was cooler under the dome, fragrant with sage and chaparral. Waiya picked up two kingly condor feathers, which he waved in the air to punctuate his speech.

"The creation story is really what a lot of it is about," he said. "Our people came here from Limuw—Santa Cruz Island." Hutash, the Earth Goddess, had beckoned the Chumash from the island to the mainland, Waiya explained, promising them a paradise for future gen-erations. She made a shimmering bridge out of a rainbow and told them to walk across it, over the sea. But she warned them not to look down. "Well, some of the people couldn't resist and they did look down," Waiya continued, "and they got dizzy and fell. As they hit the water and started to sink, Hutash asked our God, *Kakunupmawa,* our creator, our grandfather, 'Don't let them die.' Down they went and as they fell to the bottom of the ocean, their bodies started becoming silky and then these fins came out, and they surfaced and took their first breath of air. They turned into *A'lul'koy*. They became dolphins."

Waiya reached into a pouch that was slung over his shoulder. He pulled out a rattle and began to shake it and sing. The song was haunting, full of cries and long, rasping growls. I felt the skin on my arms tingle. Sumo lay down and put his head on his paws. The Chumash tongue was unlike anything I'd heard before: it was soft and hard, flowing yet crisply defined. There were sibilant sounds and shushing sounds and clicking sounds, all coming from somewhere deep in the windpipe. Waiya's words were as primal and indescribable as nature herself.

Watching him as he finished—singing the ancient language with his head thrown back, condor feathers in one hand and rattle in the other, his face pierced with bear bone—you wouldn't guess that in his thirties Waiya had been a busy building contractor, well assimilated into the SoCal world of fast food and twelve-lane highways, far detached from his native roots. It was only after a series of fateful coincidences that he'd reconnected with his lineage. One day, driving home from a construction job at Pepperdine University, he'd glanced aimlessly out his truck window just as he passed the Wishtoyo site. "I felt this *vortex*," Waiya recalled. "And I wondered, '*What is it?* Why does it feel like I'm seeing some long-distance vision?'" The site was county-owned wasteland, desolate and strewn with junk, choked by invasive plants. After that, every time Waiya saw the place it drew him nearer. He had no idea why. Later, he would learn that it had once been a thriving Chumash settlement.

That same evening, at his house, some local kids came by with flyers for a celebration at the Chumash Cultural Center. "Really?" Waiya remembers asking them. "I'm Chumash. There's a cultural center? Where is it?" They pointed to a mountain directly behind Waiya's home. This would be remarkable enough, but Waiya had moved in only a few days earlier.

He visited the center the following weekend: "And my whole life changed forever." Waiya then spent a decade immersing himself in the Chumash ways, rekindling his inner embers, apprenticing with one of the tribe's few remaining elders. When Waiya expressed frustration at how hard it was to take everything in, the elder disagreed: "No. It's unlearning what you've learned that's the hard thing."

Given the significance of the *A'lul'koy,* one of the most important Chumash ceremonies was the Dolphin Dance. Now, as an elder himself, Waiya performed it. He showed me a photo of the dance in progress: Waiya was clad in full dolphin dancer regalia, body-painted with black and white stripes and dots, wearing an elaborate feather headdress and skirt, both hands holding *wansaks,* or clapper sticks. Fire blazed in front of him. "It's a trippy dance," he said, grinning. "The clappers sound like dolphins."

Eventually Waiya returned to the land that had called to him, and was able to begin the process of reclaiming it. He and his wife, Luhui'Isha, had started by clearing the space for a single *ap,* twelve years ago. Soon, they were spending days and weeks there. They went to sleep under the constellations and awoke with the sunrise. They watched eagles sailing overhead and dolphins gamboling offshore. "Of course the construction stuff was lacking," Waiya said, "but I didn't want to do it anymore." His voice dropped, and he spoke in an urgent tone: "Because I'd seen ten years go by really fast, doing bids and giving money to taxes and dealing with my employees—and this is every day. And I wondered, '*What is life about?* It can't be about just living to work.'"

These days, Wishtoyo was a hub of Chumash ritual. Waiya and Luhui'Isha hosted groups of all ages, teaching their people's traditions, honoring solstices, and holding retreats. One of their main audiences was schoolchildren. "I can put fifty kids in here," Waiya said, gesturing around the *ap*. "I have a PowerPoint presentation about marine life. I'll tell them all about the ocean, about ecosystems and endangered species, all these different ways of looking at the environment and stewardship and balance and understanding, not just the science and the laws, but how it's a part of you and you're a part of it. And they're watching and smelling and seeing and hearing the songs and stories of a people who were living the heartbeat of life itself, not just this fabricated system that's telling you that your freedom is this document, not your birthright to have healthy land, water, and air. *That's* your true freedom. Fight for *that*. Be the voice of that which can't speak for itself. Because that's yours too." He laughed. "So I'm building little advocates."

Once he got rolling, Waiya's voice rose and fell with the cadence of a chant or an incantation, picking up emotional speed too. He was a natural orator. It was easy to envision a bunch of kids sitting in here, transfixed. "I tell them, 'If you love that ocean, that whale, that dolphin, that forest, that river, that bear, those eagles—all those beautiful things that are part of the world,'" Waiya said, "'then you've got it. It's already in you.'"

One of the most admirable things about Waiya was how he applied the Chumash teachings in ways his ancestors had never dreamed of. Teaming with a talented young lawyer, Jason Weiner, and ocean advocacy groups like Surfrider, the Natural Resources Defense Council, and Robert F. Kennedy Jr.'s international Waterkeeper Alliance, Waiya proceeded into state and federal courtrooms. Armed with a fine-tuned understanding of Native American rights and cultural protections, he set out to assert them. "We can't be the people that we used to be," Waiya said. "That's past. But we *are* the people who are going to make the future. And we have an obligation to protect our home."

The list of eco-triumphs that he and his partners had under their belts ran dozens of pages long. They'd sued private and public enterprise, cities and counties, the state of California and the federal government, to stop reef-killing discharges of toxic effluents from a wide array of sources; they forced polluters into court and made them clean up their acts to prevent further damage. They challenged a megadevelopment that would have erected twenty thousand new homes, bulldozing Chumash burial sites and destroying a fragile watershed—a battle that continues. They petitioned and filed official complaints at a relentless pace.

Recently, they had helped derail the Pacific Gas & Electric Company's plan to blast eighteen airguns—250 decibels each, firing every fifteen seconds for seventeen days—across a 130-square-mile stretch of coastal waters, a sonic onslaught that would have harmed vast numbers of dolphins, whales, porpoises, sea otters, seals, turtles, squid, and fish (not to mention surfers, swimmers, divers—anyone who happened to be within earshot of such crippling emissions of noise). PG&E argued that it needed to test the seismic durability of the seafloor (although comprehensive studies already existed), and that the benefit to its Diablo Canyon nuclear power plant outweighed the "unavoidable adverse impacts." At the hearing, Waiya gave an impassioned speech that was later singled out as the day's most moving testimony. The dolphins, he told the California Coastal Commission, were not mere statistics on a chart—they were family. "They mourn the death of their loved ones just like you do," Waiya reminded them. "No one has the permit to take lives. This is what our ancestors say." In the end, the air guns lost. The vote was unanimous against them.

Currently, Wishtoyo is working with government agencies to create a network of marine protected areas along the California coast, patches of ocean that will be fully restored and allowed to flourish as they did thousands of years ago. "We gotta do the right thing somewhere," Waiya said, as we exited the *ap* and walked to the edge of the bluff. "Our oceans are the bloodline of our world."

The sun was slinking down, flattening the Pacific into a zinc-dark slate, while a ghost moon rose above it like a pale shadow. Behind us, traffic streamed along Highway 1, but all I could hear was the sound of the waves. Below us, silhouetted in the amber light, two surfers waited for the day's last sets, floating among soft rafts of kelp. I wished I was out there with them, that I could take off my shoes and stay here longer, while my e-mails and to-do lists stayed somewhere else, at least for a little while.

I mentioned this to Waiya, and he ran with it. "We're living in a computer screen!" he said. "All this technology is gonna be our demise. And these health ailments that we're getting, diabetes and obesity, and now our social skills are being threatened because we're texting and we're e-mailing and we're not even talking anymore." He inhaled deeply. "I want to see your eyes, I want to hear the sound of your voice, I want to *understand* what you're saying. Because you can't put a breath in a sentence."

In his own badass, eco-warrior, poetry-slam way, I realized, Waiya's was a voice that combined ageless wisdom with the quicksilver tempo of modern life. Wishtoyo wasn't some quaint antique village—it was a reminder to breathe. It posed the question: When something precious is at stake, why not slow down and consider the options, not just for yourself but for, as Waiya put it, "a future that you will never see"? His words struck at the heart of a paradox that had been nagging at me: In a time when we've never had more knowledge to inform our actions, we've never been more heedless. *You are living for the moment,* Wishtoyo whispered, *while everything else around you is living in geological time.*

"We've lost our way," Waiya said, matter-of-factly. "We're an insatiable people who always want more. Dolphins and whales have their own language that they've had for millions of years, and us, we keep adding words to the dictionary because we're never satisfied. So where are we going so fast?" He laughed, and swept his condor feathers across Wishtoyo, as though blessing it. "Our stories are part of our world, and our dance, and the way we relate, and that meditation, that therapy, that ceremony, that medicine, that magic we hear when we come together. It's bitchin'! That's why I built this place."

Deep beneath the waves, the Chumash believed, there is a crystal house guarded by mermaids, frequented by the *A'lul'koy* and the swordfish people, the *Elye'wun*. Their stories also spoke of seven crystal rods buried under the ocean, vibrating in tune with the elements. "Crystal is like water," Waiya told me. "It holds the primary colors of life. It's like the diversity of humanity: We come from all over the spectrum. And in that, rainbows are born."

In the Chumash tradition, as in so many others, the dolphins were the keyholders to unknown realms, the emissaries between their element and our own. "The *A'lul'koy* represents the west," Waiya said, nodding at the horizon, "where the day ends and your dreams begin, where the land and the ocean meet. It's transition, where our people exit this world for the spirit world. And one day will be our last sunset and we, too, will transition from this life to the next." As if to illustrate his point, the moon suddenly grew brighter, taking its place on the stage.

"There's another dimension out there," I said, pointing to the ocean.

"Oh yeah," Waiya said, bowing his head. "It's real."

CHAPTER 8

THE WORLD'S END

Even from ten thousand feet in the air, the Solomon Islands looked foreboding. The ocean had a strange, Hadean stillness, so glassy the clouds reflected off it as though in a mirror. Nothing was moving; there wasn't a whitecap or even a ripple as far as I could see. Below the surface, streaks of reef ran like veins. On the horizon, the green-black peaks of Guadalcanal reared up, a jungly netherworld that saw some of World War II's most vicious fighting. Sunken warships from seven major naval battles rest in these waters, so many skeleton vessels that the area is known as Ironbottom Sound. All over the Solomons, rusted-out, bombed-out, cast-off military equipment litters the landscape, forgotten but not lost. While the rest of the world carried on and left this corner behind, it devolved into a nearly failed state. The government teeters precariously; tourism is close to nonexistent. Beheadings occasionally occur. Rare strains of malaria thrive. I pressed my forehead against the window of the Air Niugini plane and wondered what would be waiting for me on the ground.

Only a week ago, I had no plans to come here. I'd hoped that O'Barry would eventually return to the Solomons, and I thought if I ever made my way to these islands, it would be with him. Given the chaos here, I wasn't in a hurry to visit. But then the news broke that a remote village called Fanalei had killed nearly a thousand dolphins in two days, followed by another three or four hundred dolphins the next day, and that the villagers had vowed to continue killing as many dolphins as they possibly could until they were paid a huge amount of cash. It was a hostage situation, tense and gnarly and haywire, and ironically, one that had begun with the best of intentions.

In 2010, O'Barry and Mark Berman, along with Lawrence Makili, Earth Island's local director, had ventured to Fanalei and two other villages, Walande and Bita'ama, and offered a proposal: If they would stop hunting dolphins, their communities would receive financial support. Money would become available to build schools, create sustainable businesses, and shore up houses. After much communal discussion, all three villages accepted the deal. Earth Island then began to release the funds, large sums entrusted to village elders. The process was complicated because the Solomons culture involves a system called *wantok,* which requires that families, villages, and other interested parties who are yoked together by blood and obligation share everything. The spoils are owed to everyone. It's the Melanesian version of Social Security. In a tribal way, *wantok* offers protection, but because everybody's hands are out all the time, it can also result in nepotism and barely disguised looting.

In Bita'ama and Walande, things went smoothly. The money was transferred and used as intended; the dolphin hunting halted. In Fanalei, however, things went awry. After the first payments were wired they promptly disappeared, siphoned off by a splinter clan of Fanalei associates who lived in Honiara, Guadalcanal's capital city. The people who were supposed to distribute the village's cash claimed they'd never received it, despite records proving it had been sent. Expressing a typically Solomon Islands viewpoint, one Fanalei elder explained, "It is far better for us to steal our own money than have a total stranger misuse the money." When the agreement between Earth Island and Fanalei lapsed, it was not renewed.

Meanwhile, in the village of Fanalei itself—an outpost near the southern tip of Malaita, an island located sixty miles by sea from Guadalcanal—the locals who hadn't received any money looked for someone to blame. Everyone was enraged at everyone. Wilson Fileil, the chief who had brokered the deal, was forced out. The villagers, he told the *Solomon Star* newspaper, had embarked on "a killing spree."

When pods swam by their island recently, the men pushed off in their dugout canoes and captured about 900 bottlenose, spinner, and spotted dolphins, including 240 calves. Video of the hunts ended up on Al Jazeera, revealing the scene in gruesome detail. The villagers' hunting method is primitive—banging stones underwater to disorient the dolphins and then running them to the beach—but deadly effective. Once the pods were driven into the mangrove shallows, the Fanalei women waded into the water too, wrestling dolphins into canoes, dragging them onto shore by their tails, grabbing them by their beaks and slinging them over their backs. While men whacked at the thrashing animals with machetes and women harvested the teeth, children played with headless dolphin carcasses, lolling in pools of blood.

The next day the villagers hunted again, rounding up hundreds more, and the frenzied killing continued, with a tide of dead dolphins drawing flies on the beach. Far from subsistence hunting or cultural tradition, Fanalei was engaged in a massacre intended to provoke attention. INTERNATIONAL OUTRAGE OVER SOLOMON ISLANDS' DOLPHIN SLAUGHTER, read *ABC News Australia*'s headline. DOLPHIN WAR

HEATS UP, the *Solomon Star* reported. *The Guardian* also weighed in: SOLOMON ISLANDS VILLAGERS KILL 900 DOLPHINS IN CONSERVATION DISPUTE.

≈

To grasp the pitch of local hostilities, you have to consider the Solomon Islands' history, and it isn't a pretty tale. The Solomons existed apart from civilization until the nineteenth and early twentieth centuries arrived, bringing cruelty: European traders enslaved thousands of natives and forced them into labor on sugar plantations. Britain claimed the place, but it was on the far fringes of the empire. These islands played a key role during World War II: nearly forty thousand Allied and Japanese troops lost their lives here, and the fighting also took a steep native toll. Discarded live ordnance is still scattered across the landscape; in recent years, homegrown militias have dug up old shells and used them against their enemies. Vintage munitions are also picked up by fishermen and exploded on reefs. When the dead fish bob to the surface, they are easily collected; the coral is reduced to rubble.

In 1978, the islands gained independence and struggled to find their footing amid clan violence. (It didn't help the Solomons' sense of unity that ninety different languages were spoken throughout the country, by a populace strung across a vast archipelago.) Logging companies from Asia poured in, bribed officials, and razed old-growth forests. Ethnic clashes and weak government led to a savage civil war from 1998 to 2003, a period known as "the Tensions." It was a lawless time, people turning on one another with automatic weapons, World War II detritus, machetes, knives, and anything else they could get their hands on. It took an Australian-led, multinational peacekeeping force (RAMSI: Regional Assistance Mission to Solomon Islands) to put a lid on the marauding; the soldiers are still in residence today.

Unlike Japan, the Solomon Islands really do have a tradition of

Spinner dolphins off the Kona coast on the Big Island, Hawaii.

Spotted dolphins: known affectionately as "the Bikers."

Joan Ocean,
among the cetaceans.

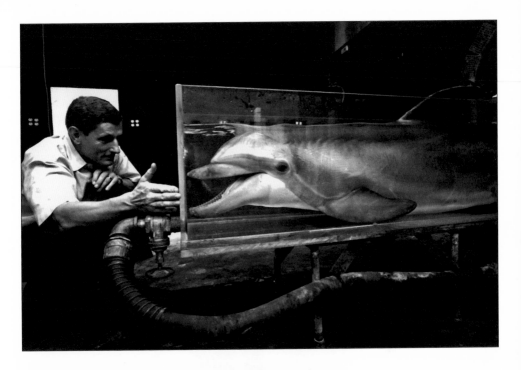

ABOVE: Dr. John Lilly with Elvar, a "bold and pushy" dolphin.
© *Flip Schulke*

LEFT: Marine Studios, St. Augustine, Florida.

A pilot whale called Moby performs in St. Augustine . . .

. . . along with a Risso's dolphin, name unknown.

"The Most Loyal Animal on the Planet": Fungie the dolphin.

LEFT: One of the five female bottlenoses who played Flipper, the dolphin television and movie star.

BELOW: Ric O'Barry, Flipper's former trainer, now devotes his life to dolphin welfare. He is shown with Angel, an albino bottlenose calf whose mother and podmates were killed in the annual drive hunt in Taiji, Japan.

LEFT: Taiji's dolphin-hunting boats leaving the harbor at dawn.

BELOW: A bottlenose pod corralled in the cove.

BOTTOM: Dolphins being selected for marine park export.

Top and above left: Dolphin slaughters under way at the cove.

Above right: An exhibit at the Taiji Whale Museum.

Left: The dolphin pens at Taiji.

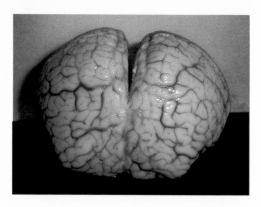

More than one way to be smart: Lori Marino, scientist-advocate (LEFT); the exceptional brain of the common dolphin (ABOVE).

BELOW: Amazon river botos, or pink dolphins.

LEFT: Fanalei, a dolphin-hunting village in the Solomon Islands, embarked on a "killing spree."

BELOW: Tribesmen release dolphins in the Malaitan village of Bita'ama.

ABOVE: Bottlenose dolphins languish in the sea pen at Honiara's Kokonut Café.

LEFT: Lawrence Makili.

RIGHT: Ric O'Barry with village chief and dolphin priest Emmanuel Tigi, negotiating a moratorium on the traditional practice of hunting dolphins for their teeth.

ABOVE: A false killer whale in motion, and (BELOW) on the hunt. RIGHT: Robin Baird and Daniel Webster survey the sea.

RIGHT: A melon-head pod in the waters around Hawaii.

LEFT: Kiska, a lone orca, circles her tank at Marineland Canada. The park is the site of frequent and impassioned protests.

RIGHT: A few of the many beluga whales kept at Marineland Canada.

BELOW: The lost city of Akrotiri, a site that has yielded many mysterious dolphin artworks.

ABOVE: Minoan dolphin fresco in the Queen's Megaron, Palace of Knossos, Crete.

BELOW: Minoan marine-style artworks found in the ashes of Thera.

Rough-toothed dolphins, fin to fin.

hunting dolphins that goes back centuries. In the past these hunts were sacred events, called by dolphin priests. They happened seasonally, and were modest in their take. Only spinner and spotted dolphins could be killed, as few as possible to serve the village's needs; there were prayers and rituals to honor the dolphins who gave their lives. Then, it was a deeply spiritual endeavor; now, like so many pursuits, it's mostly about cash. Dolphin teeth are prized as a currency, used in rural commerce. They are required, for instance, to buy cigarettes, a pig, or a bride—a woman costs at least a thousand teeth. During ceremonies in dolphin-hunting communities, both men and women will be decked out in dolphin-tooth necklaces, earrings, headdresses, belts: a lone person might be wearing twenty dolphins. Each dolphin tooth is worth between fifty cents and one dollar, depending on its size and quality. The more teeth a family displays, the higher its social status.

Into this combustible mix came a Zippo lighter, arriving at the peak of the Tensions in 2002. Christopher Porter, a thirty-two-year-old former marine mammal trainer from British Columbia, Canada, stepped off the plane in Honiara and made his way to Fanalei and Adagege, another dolphin-hunting enclave on Malaita. Porter, a burly guy with a surplus of nerve, had a vision that he expressed to the villagers in pidgin, the Solomons' lingua franca mash-up of English and Melanesian. The dolphins they were butchering for their meat and teeth, Porter told them, were extremely valuable in the outside world. If the villagers helped him catch them, they would receive great benefits. Porter wanted to sell dolphins from these waters, and then use the money from those transactions to build a luxury resort nearby where guests could "get closer to dolphins than they ever dreamed." He seemed unconcerned that a country overrun by warlords was generally not a big draw for tourists.

The Malaitans listened. If they had a talent, it was for snagging pods of dolphins. Their villages had nothing, and Porter was offering jobs, boats, cash. The tattered government granted Porter a

100-dolphin export permit and a lease on forty-acre Gavutu Island, two hours from Honiara by boat. Gavutu had a small harbor and was used by World War II Japanese forces as a seaplane base until the U.S. Marines wrested control in a mean fight on August 8, 1942.

Porter partnered with a Malaitan chief named Robert Satu, and almost immediately ninety-four dolphins were hauled in and penned up in Gavutu, and at a grotty marina in Honiara. In July 2003, the two men exported twenty-eight bottlenoses to Parque Nizuc, a swim-with-dolphins facility in Cancún, Mexico. (Like most countries, Mexico has banned the capture of dolphins within its own waters.) One of the dolphins died upon arrival; the surviving twenty-seven swam in tight circles emitting high-pitched screams for several days. Within eighteen months, another nine were gone.

Porter and Satu's dolphin exploits, once they became known, drew international condemnation. When journalists tried to investigate the Honiara dolphin pens, they were met by armed Solomon Islanders—gang members hired by Satu, who looked on with a satisfied smile as tribesmen slapped the news cameras away. A reporter from the *Sydney Morning Herald* was hit in the head with a concrete block. Chief Satu was a small, flinty man who looked wizened beyond his fifty-one years; dolphin-tooth necklaces crisscrossed his chest like bandoliers. This business of shipping dolphins to marine parks was massively profitable, he said: "It's big—bigger than gold or logging." Satu mused about the possibility of every village having its own "dolphin farm." "We've already created the market," he said. "They could just follow."

Not long after the Cancún export, O'Barry and Berman flew to Honiara. O'Barry viewed video of the remaining dolphins in Porter's pens, a number that was originally estimated at twenty-seven but had shrunk with each passing week; the remaining animals were so malnourished and dehydrated that some had developed a condition called "peanut head," their skulls showing clearly beneath their skin. He discovered that some of Porter's bottlenoses had died after being fed rotten fish; another environmental group reported that a dolphin had been eaten by a crocodile. Without notice, O'Barry showed up at Gavutu; his boat was chased away by testy guards. Later, Porter would admit that all of his dolphins had vanished while he was out of the country, fund-raising. He provided no explanation for what had happened to them.

Porter seemed to court attention, but it was clear the antagonism rattled him. "Like the rest of the world, just blame Chris Porter," he complained. "I'm the monster in activists' heads. Everyone thinks I'm the worst man on Earth for dolphins. I'm the one who catches them all and I'm the greedy one and I'm the exploiter." According to Porter, the opposite was true. "I love animals," he said. "I cried during *Old Yeller*." Few people, however, saw him as a friend to wildlife. Porter was dubbed "the Darth Vader of Dolphins," a nickname that stuck.

Porter explained to the media that he'd chosen the Solomon Islands for his operation *because* of the country's dolphin-tooth fetish. If they were decapitating so many of the animals, he reasoned, it seemed unlikely the islanders would mind exporting some too, particularly if they profited from the sales. But one local man, Lawrence Makili, did mind. He minded a lot.

Makili was born on a remote atoll called Ontong Java, one of the farthest-flung bits of the Solomons, technically part of Malaita even though it is three hundred miles away. "I came up from nowhere," he'd said, describing his roots. He had four brothers and two sisters, many nieces and nephews and cousins, two adult sons, and a fiancée. About half of his family lived in Honiara, while the rest had stayed on the atoll. In his youth, Makili made his way to Guadalcanal to attend college, at an institution run by the Anglican Church, but he was expelled during his third year after protesting the misuse of school funds.

Above anyone else in his country, Makili was the face of opposition to the Solomons' dolphin trade. He was a devout activist, confronting rapacious animal traffickers and slimy officials and despoilers of the natural world. Before he worked with Earth Island, Makili had worked with Greenpeace. He'd fought illegal logging—an issue that had gotten many dissenters murdered—and nuclear testing in the South Pacific, and wildlife poaching, and all manner of dirty dealings in the Solomons. His willingness to speak out had made him a target: in 2008, Makili barely survived an assassination attempt. In a land of tough people, he was one of the toughest. But he was one man against an avalanche of corruption, self-interest, and money, in a place where survival was far from assured. The odds were not in his favor. Makili had friends, but he also had plenty of enemies.

≈≈

When I heard about the Fanalei dolphin slaughter, I called Mark Palmer at his office in Berkeley. "It's a long, complicated story," Palmer said, with a sigh. The money had been stolen by the village's urban faction, he confirmed, and what's more, there was suspicion that dolphin traffickers had encouraged the theft. Since 2003, dolphins from the Solomons had been shipped not only to Mexico, but also to Dubai, China, and the Philippines, despite a global outcry against the practice. It was in the traffickers' interest for the hunts to continue, Palmer explained, because then they could export the animals with less controversy, by claiming they were "saving" them.

"Well," I said, after listening to Palmer. "I think I should go there right now."

"Ahhh, that would be exceedingly dangerous," he said, and then laughed sharply, as though astonished by the suggestion. "You'd need an armed guard for sure." Heading to the Solomons was a terrible idea, he stressed. Things were far too volatile right now. Even Makili was laying low these days. "Being seen with Lawrence would taint you," Palmer warned. "And they've tried to kill him before."

To poke around the country on the sensitive subject of dolphins would be like pressing on an exposed nerve, but that was the point. Where humans meet dolphins the result is a spectrum of extremes, and only in the Solomon Islands was every last one of them on display: dolphins were worshipped and abused, revered and gutted, valued and discarded. They were considered mystical, and they were used to buy women. While there were many places I would have preferred to visit, there was no place I needed to witness more. I let Palmer try to dissuade me for a bit longer, and then I hung up and began to search for flights online.

I quickly discovered that planning anything in the Solomons demands persistence. The hotels answered their phones only rarely; flights into the country were sporadic. I made e-mail contact with Makili: he instructed me to rent a car, but I was quoted a price of $4,000 U.S. a week for a rusted-out subcompact. "Can you help me find a driver or a guide or a bodyguard or—something?" I e-mailed back. "I can recommend you a driver," Makili replied, "but what is he going to drive if you aren't hiring a car? Please inform me properly." I wrote back, but Makili had vanished. I didn't hear from him for days. Finally he resurfaced, ignored every one of my desperate queries, and told me flat-out to settle down. "Safety now is a bit risky," he wrote. "But I can handle it. It's okay, this is my country."

Somehow, I'd sort of arranged that someone might be at the Honiara airport to meet me and drive me to my hotel, the Kitano Mendana. When I spoke to Berman about my trip he warned that I'd be followed in the city, and that I should never venture anywhere alone. "That's the way it is there, I'm afraid," he said. "Please believe me." My fears weren't calmed by the eight-foot statue of a Malaitan warrior, lunging forward with a spear in one hand and a club in the other, that greeted arrivals in the one-room terminal.

Rolling my bag out to the road, I noticed a smiling man in crisp slacks and a white button-down shirt looking at me. "Mendana?" he said, pointing to an unmarked van. "Yes," I said. "Are you from the hotel?" He didn't answer, but reached for my suitcase and flung it into the back. "Come," he said. Lacking other options, I got in.

We swung onto the main road, possibly the only road, two lanes in each direction split by a center median. It was high noon and the sun was hard and bright. Women trudged along the dusty shoulder carrying umbrellas for shade; a man walked by with a live pig slung across his back. Despite the poverty in evidence, the landscape was lush. Palm trees towered; grass grew exuberantly. The few buildings I could see were squat and crudely built, thatched or tin-roofed, wrapped with fencing and barbed wire. "How far is the hotel?" I asked, mainly to confirm we were actually headed there. The man glanced at me in the rearview mirror. "You are Susan," he said, with a nod. "I have talk of you."

His name was Albert, he told me, and he was from Savo, a volcanic island just north of Guadalcanal. He was kind and educated and chatty, taking the opportunity to practice his English. The main thing he wanted to discuss, it turned out, was dolphins. "The people should cooperate to look after the dolphins," he said, with conviction. "The dolphin is very like myself and yourself, Susan."

I wasn't expecting to dive into the nation's touchiest topic quite so instantly, and I wondered if it might be some sort of a setup. I had no sense of why the driver would connect me with the dolphin controversy, and this was worrying. For safety reasons, I needed to keep a low profile. Perhaps the average Solomon citizen, like Albert, wanted to protect dolphins, but to the villagers who were hunting the animals, or the individuals who were in on the live capture trade, anyone questioning either practice was an unwelcome guest. During O'Barry's last visit here, a prominent Honiara businessman and dolphin dealer had sent a posse to attack him right in his hotel lobby. The list of beneficiaries from dolphin trafficking reportedly reached high in the government; its official response to Makili's abduction and beating was to criticize him for, in the words of the fisheries minister, "trying to disrupt the country's new multi-million dollar industry."

Then, of course, there was the dolphin-based currency system. "It's part of our culture, taking the teeth," Albert confirmed, as though attempting to teach a crash course in Solomon Island Dolphin Studies before we arrived at the hotel. "When you make a bride price for marriage. It's very strong in Malaita province." There, the rule was simple: No dolphin teeth, no wife. "They killed more than a thousand-plus dolphins last week," Albert said, shaking his head. "That's not good."

I didn't know what to say. We had stopped at a light, pulling up next to a cement planter painted in a tribal motif, depicting a man straddling a dolphin and bending it painfully backward. The man had the dolphin's beak in his mouth, as though devouring the animal headfirst. "I'm honest to tell you, Susan," Albert continued, "we always swim on the sea, and the very young dolphins come and just play. It's beautiful!" He smiled into the rearview mirror. "I think we have a dolphin cave," he concluded, describing how the pods seem to materialize out of Savo itself, and then mysteriously disappear back into its bowels. As he turned into the hotel, an A-frame building shrouded in palm trees, he wound up with one last pitch for his home island: "We would never do any harm to the dolphins, never. Never!"

≈

The hotel was relatively modern, and located on what passed for a beach in Honiara, a muddy strip confettied with plastic. A man with ritual scars across his face checked me in with an absence of chitchat. My room was clean and equipped with double door locks, so after I'd shoved my suitcase into the closet, popped my malaria pill, and washed it down with a swig of vodka, I crashed for a nap. I'd been traveling for thirty-two hours. It took only seconds for me to pass out in exhaustion.

I woke to the phone ringing insistently; someone was calling from the lobby. This was unexpected. Who even knew that I was here? Makili had instructed me to buy a local cell phone, then text him when I had it. I'd figured I would do this in the morning, and then meet him in some out-of-the-way place, as instructed by absolutely everyone. It had been repeatedly emphasized to me: being seen with Makili would draw the kind of attention I didn't want. I picked up the phone. "Hello?"

It was the scar-faced man at reception. "Lawrence Makili here to see you."

Within seconds, I heard a knock at my door. I opened it and there was Makili, familiar because I'd seen photos of him, including some taken in the hospital after his attack, with both eyes swollen shut and both arms encased in casts. In person, as in pictures, Makili was not someone who would fade into a crowd. He was a substantial guy, solid as a barrel, with a corona of dreadlocks that fell past his hips. His face was kind and ferocious at once, framed by a mustache and beard that were just beginning to gray. Makili was only forty-two, but he had the careworn air of an older man. He wore slouchy pants and battered leather loafers and a T-shirt with a drawing of a crying dolphin on it.

It quickly became clear that Makili didn't share anyone else's concern about keeping out of view. When I asked if it was okay for me to be seen hanging around with him, he let loose with a belly laugh, as though this were the most ridiculous question in the world. All the mutinous strife and ugliness, the accusations and counteraccusations, the lunacy of his country's recent history with dolphins and his dicey role as the sole resident to stand up to it—all of this had left him undaunted. Makili seemed frustrated, perhaps, pissed off for sure, but not intimidated. He sat down as though ready to talk, but he was fidgety and uncomfortable in the wicker chair. Later, I would come to recognize this body language as a cue that Makili wanted a cigarette

(Marlboro Blues) or a SolBrew (the national beer), or a mouthful of betel quid (a psychoactive nut chewed for its stimulant buzz), but at the time I chalked up his restlessness to angst about the Fanalei situation.

"How many dolphins do you think they've killed?" I asked.

"Thousand plus," he answered quickly. "They hunted today again, seventy more." One of the methods the village used to lure more dolphins in, Makili explained, was to keep a small group alive in a sea pen where they would swim around making distress calls, "so the others in the pod will come back." He sighed.

"Can we go to Fanalei?" I asked. I knew it was a difficult trip, two days by boat and on dirt roads overland. But if Makili was up for it, so was I. "Ah, I'll have to talk to Chief Willy," he said. "Fanalei is a hard place." Chief Willy, or Wilson Fileil, was the man who had been scapegoated after the Earth Island money went missing. I would meet him tomorrow, Makili told me, along with the chief from Bita'ama, Emmanuel Tigi. Both chiefs were in Honiara at the moment; Willy, in fact, was unable to return to his village until the matter was sorted out. "A bunch of ignorance!" Makili said, with disdain. *"Wantoks."*

"Well," I said, still pushing for a road trip. "How about Gavutu? Can we go there?"

Makili smirked. "Oh, we'll get there."

"Will we go by boat?"

"Yeah, by boat," he said, and then burst out laughing. I wasn't sure what was quite so hilarious and at the moment didn't want to inquire. But I did want to visit Porter's encampment, which he had named "Dolphins Paradise," even though I knew the Canadian wasn't there right now. Online, I had watched a promo video set to throbbing calypso music, featuring Porter, in surf trunks and a dolphin-tooth necklace, giving a tour of his imagined resort. "Currently, we are finishing renovations to what will be one of the most unique bars in the South Pacific, overlooking our dolphin lagoon system," he said, as the camera panned across a group of bottlenoses clumped in the shallows. "We're also finally completing the VIP bungalow . . . We invite everyone to come and visit and touch and feed the dolphins and learn more about them."

Makili stood up and began to pace. "You see," he said, "I'm not comfortable here, because I feel like this is a nonsmoking area."

"Do you want to go get a beer?"

"You still want to talk?" he said. "Okay, we go talk."

≈

"This is my corner," Makili said, as the decrepit taxi puttered up the steep hillside, pulling in at his house, a brown and white two-story with a flat roof, raised on pillars. Wild hibiscus silhouetted the driveway. A deck on the second floor offered a panoramic view of Honiara, the city lights hazy from the smoke of cooking fires. It was nightfall; the edges of the sky radiated cobalt blue.

His place was only five miles from the hotel, but it had taken us an hour to get here. We had stopped at a corner store so Makili could replenish his Marlboro supply, and he had run into members of his extended family, then disappeared into the store for a small eternity. Finally, he emerged and climbed into the front seat of the cab, but not before conversing in pidgin for a while with an elderly man, to whom he handed a $100 Solomon bill. "There's a boat going home tonight to my island," he told me. "I bought rice and water and milk for all my cousins. It's culture."

"Those are your *wantok*?" I said.

"Yeah, yeah."

We sat outside at a wooden dining table, accompanied by a timid gray dog. Makili lit another Marlboro. "Anton—beer!" he yelled. In a second, a handsome young man appeared, flashed a smile, and set down two frosty cans of SolBrew. Anton, Makili explained, was his elder son, named after a Yugoslavian activist he'd worked with in the field. "To your first day," he toasted. "I will tell you the real history of the Solomon Islands bullshit."

I drank to this. "So how did all the dolphin craziness begin?"

Makili settled into his chair. Back in 2002, he told me, he had been hired to develop ecotourism ideas for the country. In the process, he'd stumbled across Porter's foreign investment application to open a marine mammal education center. "Then, in the next few days, another application from Porter arrived," Makili recalled, "for a company called Marine Exports Limited. And I knew exactly what he was gonna export."

If Porter had set out to make his pitch at a vulnerable moment, he'd succeeded. "He came at the height of the Tensions," Makili said, his voice bitter. "To take advantage of the situation. There was no law and order." During the country's civil war, he explained, traditional dolphin hunting had all but stopped. It was only Fanalei that was rounding up pods: "It was a dying practice. But it started up again because Chris Porter came waving the flag of money!" As a result, Makili added, a half dozen villages had now revived their hunts, hoping to catch saleable dolphins. He took a slug of his beer and fixed me with a steely glare. "Porter is responsible for the escalation of dolphin hunting in the Solomon Islands. And he introduced the dolphin trade here."

Porter's first dolphin shipment had occurred even before the RAMSI intervention, while the national violence was still seething. Recruiting Chief Satu was a smart move, Makili said. Satu was well connected. He was a fisherman; he knew how to catch dolphins. His village was located on a stretch of sea that bottlenoses frequented. Plus, he had unlimited access to heavyweight thugs. All of these things were useful for Marine Exports' endeavors.

But some expertise was missing. It's one thing to net a bunch of bottlenoses in the Solomon Islands and keep them in a sea pen, hire villagers to fish from their dugout canoes for the eighty kilos of food that each dolphin requires per day—and another entirely to install the animals in a marine park halfway around the world. For Porter's next dolphin delivery, another partner joined him: Ocean Embassy, a group of ex-SeaWorld employees who had formed a consultancy. They had the resources and veterinary know-how for international transport. Under their direction, Gavutu bristled with new buildings, and even plumbing and electricity. In October 2007, twenty-eight freshly captured bottlenoses were transferred from Porter's Dolphins Paradise to Honiara, where they were loaded onto two chartered Emirates DC-10s for a thirty-hour journey to Dubai. Their future home would be Atlantis, The Palm, a $1.5 billion resort built on an artificial island made from ninety-four million cubic meters of sand, dredged out of the Persian Gulf.

A Canadian documentary crew was able to film parts of the process, with Porter striding across the tarmac among Satu's tribesmen and throngs of police, and Ocean Embassy staff in button-down shirts and mirrored sunglasses hustling around with walkie-talkies. Each dolphin was encased in a white canvas sling, with holes cut in the sides to allow for pectoral fins. The animals were draped with wet towels to keep their skin from burning in the sun or drying out, and they thrashed and cried shrilly. Nothing in their short lives had prepared them for this day, for being hauled out of the water and loaded onto a barge, then transferred to a truck, bumped across washboard roads, and forklifted into the planes; for the handling and the pressure changes and the grinding whine of jet engines and the crush of humans around them. As the trucks pulled up to the hulking DC-10s, security guards stretched a blue tarp across the cargo bays, blocking the view. No one involved in this operation—the largest, most expensive, most audacious dolphin sale ever undertaken—seemed too eager to have it memorialized on film.

The outrage over this second shipment was even louder. NO MERCY, screamed the *Solomon Star*'s headline the next morning, a photograph of a dolphin immobilized in its sling spread across the front page. "This is appalling," declared Australia and New Zealand, officially. In the press, Porter was referred to as a "dolphin-napper," and a "dolphin slave trader," and compared to Colonel Kurtz in Joseph Conrad's *Heart of Darkness*. To activists' dismay, no confirmation was provided that the bottlenoses had survived the trip. Upon arrival, the dolphins were swept into the Atlantis and never publicly accounted for, just another twenty-eight of the 65,000 marine animals collected for the hotel, including a fourteen-foot whale shark who served as the center-piece of the lobby aquarium.

Not everyone agreed this was a nauseating turn of events. "It is not only a great day for [Marine Exports Limited]," Porter announced in a press release, "it is also a great day for the Solomon Islands." The country's fisheries minister bragged that each Dubai-bound bottle-nose had sold for $200,000 U.S., with the government receiving a 25 percent export tax. Ocean Embassy's vice president of international operations, Ted Turner, complained to the film crew that environmen-tal groups had unfairly maligned his company's efforts: "It's a shame because it leaves the general public with the impression that something wrong has happened, when this couldn't be more right," he said, add-ing: "It cost millions—literally millions—to do a transfer of this mag-nitude. And the reason that doesn't bother us is because these animals are worth it."

Turner spoke in an aggrieved, over-enunciated tone that brought to mind a fed-up boss lecturing to an especially slow employee. "These animals are generating profit," he continued, as though the dolphins had just released their quarterly sales numbers. "And I wish every spe-cie [sic] in the world would generate the level of profit these animals are for their companies. If that were the case we'd do more to protect them."

Makili, as always, was the loudest native voice. Right before the planes departed for the Middle East, he distributed photos he'd taken of three bloated, decomposing bottlenose carcasses being eaten by dogs at a dump site in Honiara. By now, Porter and Satu weren't the only dolphin traffickers in town. Their widely publicized paydays had inspired several copycat operations, all vying to sell the country's dolphins. Makili's outspoken criticism of the dolphin trade was becoming increasingly risky; people warned him he was putting his life in jeopardy. They were right.

Anton appeared with new SolBrews and I steeled myself to ask Makili about his attack. It had nearly been fatal, but I didn't know much more than that. "What happened?" I asked. Makili inhaled slowly through his teeth. "It was because of my campaign to expose the dolphin trade," he said, in a low, raw voice, as though the memory still scalded. "Because they wanted to keep it so very secret."

On the eve of the assault, a sultry August night, Makili was on his porch drinking beer—"I was drunk, you know"—when someone snuck up behind him and bashed him in the head with a bat. "I was knocked out," he recalled, "and my senses slowly came back. Then I realized that I was in the boot of a car." Though he didn't know it at that moment, Makili was in the company of no fewer than nine abductors, the gang traveling in two vehicles and headed into the hills. Jack-knifed in the back, regaining his bearings, Makili planned a surprise offensive. Feigning unconsciousness when they opened the trunk, he suddenly sprang at the nearest men. "I got two of them," he said, with a wry chortle. "I really wounded them. If it had been three or four . . . but it was nine."

Makili was dragged away and severely beaten; again, he lost consciousness. He woke up in the hospital with a broken arm, cracked ribs, fractured skull bones, a busted eye socket, split lips, and an assortment of other injuries. He was saved, the doctor told him, by a rural dweller who'd heard the commotion and shined a flashlight onto the fray. The attackers fled in their cars, leaving Makili on the ground bleeding.

"Do you have any idea who was responsible?" I asked.

"Yes," Makili said. He lit another cigarette.

No one had ever been charged with Makili's kidnapping and assault, although he'd filed a police report, identified some of his attackers, and often saw them on the street. "There was no investigation," he said, "no nothing. The case has been buried."

He stood up and shouted to someone on the driveway below. "The taxi is here to pick you up," he said. "I'll go with you, drop you at the hotel, and come back." When I said that I'd be fine on my own, Makili snorted with laughter. "I'm responsible for your safety," he said. "If they're gonna somebody kill—me, not you."

≋

HUSBAND PRIME SUSPECT IN SORCERY KILLING, read the *Solomon Star* headline the next morning. I'd picked up the paper on my way to breakfast, but soon wished I hadn't. There was no shortage of grisly news. Sorcery, apparently, was a sticky problem, not only here but in neighboring New Guinea, where twenty-nine people had just been arrested for participating in a sorcery-fueled "cannibal cult." Popular types of sorcery included, according to the article, "A'arua, Pela, Vele, black magic, green leaf and so forth." There were also stories about revenge killings, death by crocodile snatching, and rumors that a riot was brewing in Honiara.

The morning was swampily humid. I sat outside drinking my coffee and gazing at the harbor. Naked kids played in fetid water near an outtake pipe; oil drums were unloaded onto a dock. I watched a man hurl the dark contents of a bucket into the ocean, allowing it to splash onto his legs and feet. At noon, Makili was due to pick me up so I could meet the chiefs. I found him in the parking lot, pacing and smoking. We got into the taxi and drove into the heat of Honiara, the streets dense with people, Makili barking directions to the driver in pidgin. The air tasted of overripe vegetables, diesel fumes, and rotting fish.

Buses trundled by, with riders packed in so tightly some were hanging off the back. There was a grim fatigue in their faces, a major shortage of joy.

We motored up the hill, leaving the city grit behind. The two chiefs sat waiting on Makili's balcony, a dolphin species chart spread on the table before them. "This is Willy from Fanalei and Tigi from Bita'ama," Makili said, introducing us.

I recognized Tigi from a television series about the dolphin trade. His chiefdom was a clutch of nineteen villages in North Malaita, home to four thousand people. On the program, Tigi was shown negotiating an agreement with O'Barry for Bita'ama to stop hunting dolphins. In one scene, he appeared in his dolphin priest costume—a sarong, ropes of dolphin-tooth necklaces, a dolphin-tooth headdress, ceremonial armbands, and what looked like a set of raffia wings on his back—and announced that the villagers demanded $12 million U.S. a year to lay down their machetes. If the money was not paid, Tigi had bellowed, "We will slaughter the dolphins of the whole earth!"

In the end Bita'ama settled for a lesser sum, and these days Chief Tigi was a happy man, thrilled with Earth Island's contributions to his villages: fuel drums, machine parts, a portable sawmill, lumber for houses. He was a skilled spokesman, neat in Oakley blade sunglasses and a white golf shirt, outlining Bita'ama's new dolphin-friendly philosophy with a politician's air. "We normally kill and eat dolphins," he began, "but now we want dolphins to be safe." As proof, he showed me a photo of some tribesmen, standing waist-deep in the ocean snuggling dolphins in their arms like babies. When the agreement with Earth Island was signed, Tigi said, the villagers had 160 dolphins penned in their lagoon, ready for slaughter. Instead, as a gesture of good faith, they had released them.

Willy was quieter, shy even. He wasn't as sunny as Tigi, but then again, Tigi hadn't been ejected from his village—Willy had. He wore head-to-toe black: black jeans, a black Nike T-shirt, and a black bucket hat. Both men were barefoot, even though Tigi was carrying something akin to a briefcase.

Willy explained that Fanalei was much smaller than Bita'ama, with only three hundred residents. Until recently its population had numbered 1,000, but relentless sea-level rise had forced many families to relocate. Eventually, everyone would be displaced. The ruckus over the missing aid money clearly weighed on Willy's mind, but it was Tigi who most forcefully denounced it. "What they are doing is criminal activity," he said, jabbing his index finger into the air. "They steal the money and they try to cover it up by killing dolphins. That's what it is—just a cover-up."

Willy nodded. "I can't understand these people," he said, looking dejected. "They are very complicated. Last Sunday they met about you. They know you, they know you are here . . ."

"What!" I said. "Me? They know me? Why?"

"They say you should not come to Fanalei," Willy said, staring down at the table. Tigi cut in, statesmanlike, trying to smooth things over. "It's okay," he said, smiling broadly and waving his hands as though conducting an orchestra. "You tell them, 'I am a journalist. I work on an international body.' Then, no confusion! Then we will all be peaceful."

"Today they disturb me too," Willy assured me. "They are very angry with me. But they're not gonna kill somebody. They just want to—"

"We need each other," Tigi inserted. "We need working together to solve problems. This dolphin issue is affecting even the whole world!"

I was silent, digesting this, when Makili, who had been inside the house, came stomping out gripping his phone. "Fuck!" he shouted. "They just slaughtered nineteen melon-headed whales!"

Willy nodded glumly. "Yesterday they catch more than one hundred dolphins."

"Nineteen!" Makili repeated. "Those are like killer whales but they live in tropical waters. Fuck!" The three men broke into an argument in their native tongue.

Tigi turned to me. "You are the right woman to deal with this." Before I could express more alarm, Makili quelched the notion of me as any kind of intertribal mediator. "Nah," he said, and then leaned over the railing to spit out a big gob of betel nut juice. The habit was colorful, if disgusting: it produced a scarlet, tooth-staining cud. "If you are willing to talk to them," Tigi pressed, "we will resolve all things and we will get our minds together and we will tell the world."

I tried to imagine myself landing at Fanalei, stepping over the dolphin carcasses strewn up and down the beach, strolling into a village where the chief had been exiled under threat of violence, and announcing in my Canadian-accented English that I was there to help them sort everything out. Somehow, I couldn't see it.

We went inside for lunch, a spread of chicken, rice, cooked greens, and SolBrew. I was quiet because the conversation was in pidgin. After a while, I asked the chiefs about their lives. "I myself was a dolphin caller at Bita'ama," Tigi said, outlining the ritual by which the animals were summoned to their doom. He spoke fast, inserting words I couldn't recognize, but as far as I could tell, the process involved combining thunder and lightning and rainbows to create an irresistible magnetic field that pulled the dolphins toward shore. "We're born of the tribe of dolphin," Tigi said. "We have a spirit that will call a dolphin out from the whole planet. No one knows about this."

"Do you do this too?" I asked Willy.

"In Fanalei, no, we are not calling the dolphins," Willy said. "We just hunt." Long ago, there had been special dolphin ceremonies, he said, "but now Christianity is with us so we use that in the killing. We pray for dolphins to easily come, easy to see and easy to drive to the shore."

"Does the village see the hunt as a blessing?"

"Yeah," Willy said, then corrected himself. "Well, some people think it's a blessing. But for me—the dolphin, it's a creation. A creation by God. And we just kill. We kill that mammal that is created by God. For me, for my understanding, for my opinion, that's not a blessing. It's just a form of stupid." He looked at me. "It is true that dolphin hunting is our culture. But it's time to change."

Tigi nodded. "We love dolphins! We love the teeth and meat. It taste like beef. But after the agreement with Earth Island I say, 'Hey, it's a mistake to kill these—it is wiser than a human being.' So we will build houses and income generate and the dolphins will be saved."

"We can do something different," Willy agreed, "like tourism."

"The problem is that nobody wants to come here," Makili said, unconvinced. "It's a crap country."

Tigi lit up with an idea. "We will give you land," he said, pointing at me. "You can come and develop it. You bring people here. We will work on it together!"

≋

"So, are you ready for a very long trip in a very small boat?" Makili laughed hard, then bent over and spat out a stream of betel nut. All around his feet were vivid crimson splotches, a patina from years of betel nut chewing here, as though the entire country were engaged in a furious paintball match. Willy, standing beside him, grinned. We were ankle deep in garbage, waiting at the water's edge for our captain to gas up his forty-horsepower outboard engine by pouring fuel in with a plastic cup. Our boat was twelve feet long, basically a steel rowboat. It had no shade, no radio, no GPS, no lifejackets. As for oars in case the motor quit, it had one.

Vessels like these, known as "banana boats," were a main method of transport in the Solomons. That didn't mean they were safe. Behind us, the husks of other, wrecked banana boats lay tipped on their sides, sunk into a geological strata of litter. Plastic bottles and bags, crumpled containers, scraps of metal and decaying fish guts made up the top, most visible layer. Below that was a layer of muck soaked with effluents I couldn't identify and didn't want to. As I stood there, wondering what I was in for, a black flip-flop floated by. "Whatever you do, don't get into one of those banana boats!" Berman had stressed, at least twice, before I'd left.

It was one o'clock in the afternoon and we were headed to Gavutu. We'd planned to leave four hours earlier, but Makili had been AWOL all morning, not answering his phone. I figured the trip was off, but then he and Willy showed up at the hotel with a sack of betel nuts and cigarettes, bleary, but ready to go. The sky was faintly overcast; there was a light chop on the water. It didn't seem like the worst day in the world for a boat trip, but Gavutu was two hours away.

Makili steadied the boat while I got in, clutching a bag that contained my notebook, camera, and a bottle of water. Willy followed, and then Makili shoved us off and jumped in too. The captain, a sinewy guy in his forties sporting a bushy mustache, started the motor. The deckhand, a silent, beetle-browed young man, raised a flag, and then we putted away from the shore in reverse.

I settled in front on a metal bench; Makili and Willy sat in the middle. The captain was at the stern, driving the outboard. The deckhand crouched at the bow, scanning for debris: accidents from hitting drifting junk were common. After we cleared the harbor, the captain opened the throttle and I realized, with some horror, that this was the boat ride from hell.

The ocean was rougher than it looked, and the boat went airborne off the troughs, landing each time with a sharp bang. The motion was worse near the bow, where I sat. On every wave, I felt my butt lift off the bench and then slam down hard. It was like being dropped, ass-first and repeatedly, onto a steel floor. Frantically, I cast around for something to cushion my seat—a towel, a sweatshirt—but we had nothing of the kind on board. I wore shorts and a long-sleeved shirt; idiotically, I hadn't even brought a jacket. I could see Makili watching me out of the corner of his eye, wondering how I was taking the abuse. The boat bucked and whammed its way across Ironbottom Sound. *Four hours of this just might kill me,* I thought. Then, the monsoon started.

I was so wrapped up in my misery that I hadn't noticed the livid clouds on the angry horizon. We had driven for an hour and were out of sight of land when the first bolts of lightning forked from the sky and hit the water, followed by growls of thunder that could be felt in the solar plexus. Fat raindrops spattered us, becoming torrential sheets. I glanced at the captain, who had pulled on a windbreaker but looked unconcerned. Makili and Willy sat stoically in their surf trunks and T-shirts, chewing betel nut amid the deluge. Lightning flashed again. Apparently, we were not turning around.

The storm hammered us for thirty minutes then began to lift as we neared the Nggela Islands, leaving creeping tendrils of fog. The wind had ebbed and the ocean had stilled. It was warm out, thank you God; hypothermia would have been the breaking point. I wiped the water off my face and pulled out my notebook.

We steered into a maze of undulating shorelines, all of them brooding and eerie and smothered in vegetation. At first, as we glided through channels, I could see only trees. But then my eyes adjusted and I began to make out traces of habitation: a half-sunk dugout canoe pulled onto the shore, a rickety hut on stilts overhanging a misty lagoon, just a minute from being reclaimed by the forest. This was a somber landscape: there was no mistaking its haunting past.

We pulled alongside a crumbling concrete pier, and the captain cut the motor. "Where are we?" I asked Makili, because the place was deserted. "Gavutu!" he said, gesturing with both arms toward the emptiness. I looked at the shoreline, puzzled. Where were the buildings I'd seen in the documentary? The fish houses, the sleeping quarters, the tiki bar, the floating docks? Where was the VIP bungalow?

None of it was here now—and certainly, there were no dolphins. "It's all gone," Makili said, stepping out of the boat. "Ah man, it was really a great setup. But when Porter left, all the workers, they ripped everything. Do you know why?"

I took a not-so-wild guess: "Because they wanted money?"

"Yeah!" Makili guffawed. "And they took everything! Every last piece of wood! Every last nail!"

From what I could gather, Porter's last days here had been bad ones. After one or two more dolphin exports, bound for a resort in Singapore, Ocean Embassy had broken off with him and switched its allegiance to another local dolphin dealer. Although it was unclear what caused the rift, one thing was sure: Porter's last captives, a pod of seventeen bottlenoses, had been caught in the middle. Instead of shipping them to some far-flung marine park—which would have been difficult without his partners—Porter had attempted to raise money for a project he called "Free the Pod," a fuzzily detailed initiative that challenged activists to buy from Porter the seventeen dolphins that Porter himself had captured, in effect paying a ransom for their freedom.

In the end, the campaign did not succeed and all of the dolphins died. Porter was left with debts and no way forward and he'd recently returned to Canada. It was at that point, I supposed, that Gavutu had been ransacked.

I climbed out of the boat and walked down the pier. Rain dripped from the palms onto a carpet of fallen leaves and branches. Cement slabs spiked with broken rebar stuck up here and there and some junk lay about, but Makili was right: every scrap of Dolphins Paradise was gone. I did a lap of the premises, searching for signs of its former ambitions. There were none. Even the lagoon was empty: I stared into the water and couldn't spot a single fish. All I could make out below the surface were broken pilings, snarls of wire and rusted chains. Everything was still, even the birds. Gavutu would be known throughout history for its small, nasty part in World War II, but on this drippy day it played its latest role to the hilt. It was a sad and spooky dolphin ghost town.

≈

For the next few days I stayed on my feet as much as possible, nursing the grapefruit-size, blackberry-colored bruises on my ass, souvenirs of my banana boat trip. I became bolder, too, venturing to the market and a nearby expat café, the Lime Lounge, which catered to aid workers and entrepreneurs bravely trying to do business here. That was how I found Anthony Turner, a rugged Australian who had sailed to the Solomons on his boat, a fifty-five-foot catamaran that he'd built by hand.

Turner, an avid surfer, had guessed the country's untrammeled coastlines might be a wave hunter's paradise, so he'd gathered a crew and set out on a reconnaissance mission. By the time I met him, he'd been meandering through these islands for two months. He found great waves all right, but his dream of running surf charters here had been dashed. Over espresso, Turner told me how he had dodged thieves, hijackers, shakedowns, tsunamis, and saltwater crocodiles; he'd met another sailor who had his fingers chopped off by tribesmen, and a group of do-gooding dentists, here to fix villagers' teeth for free, who were robbed of all their equipment. Fanalei's dolphin slaughter, he said, had extinguished any last hopes of establishing an ecotourism business: "That was the nail in the coffin." At the moment, Turner was in Honiara awaiting a visa for one of his crew so they could sail to the Philippines or Indonesia—anywhere but here. "Life's too short to sleep with a knife," he said.

Given Turner's willingness to visit inhospitable places, I decided to ask him for a favor. I wanted to visit the Kokonut Café, a harborfront bar, but I was hesitant to go alone: along with SolBrew, egg sandwiches, and unfiltered cigarettes, the bar's owner, a local man named

Francis Chow, sold dolphins, keeping the animals on display in squalid conditions. Chow was affluent, with close ties to the government: at the Kokonut Café's launch celebration the previous year, the country's prime minister had cut the ceremonial ribbon. Chow made no secret of his participation in the dolphin trade, and was known to be hostile to those who opposed it. If activists wanted to stop his dolphin exports, Chow told an Australian reporter, they should be prepared to give him $10 million.

Turner knew the joint, and winced as he described the dolphins' bleak holding pen: "They're in an institution. It's just imprisonment. That's all it is." He didn't think anyone would threaten us if we stopped in, provided we didn't pull out a camera. Most of the Kokonut Café's patrons, he told me, were zonked out of their minds on kava, a psychoactive root that is brewed into a drink.

We decided to go in the early evening. Turner picked me up in the dinghy he used to shuttle to and from shore, running the inflatable craft up on the beach. I hopped in; from the hotel, it was only a two-minute ride. The Kokonut Café's bar was built on a cement jetty, next to a dock where we could tie up. Far from being "an amazing project," to quote the Solomons' minister of culture and tourism, Chow's establishment was a patchwork of slapped-together structures with a sea pen gouged in the middle. Ratty wire fencing stuck up eight feet above the surface, buckled in places and gooey with scum at the waterline. Piles of rocks, hunks of concrete, discarded construction materials, and half-finished buildings added to the junkyard ambience. Music blared from loudspeakers, bouncing across the water. If there was a more wretched place for a dolphin to live, I couldn't imagine it.

As Turner secured the dinghy, a man sidled up next to us, glaring through bloodshot eyes. Without breaking his gaze he spat betel nut into the water. His lips were stained reddish-brown, as though he'd been drinking muddy blood, and his teeth were almost black. More surly men, maybe a dozen of them, sprawled in plastic chairs at tables arranged haphazardly under a tin canopy. Crumpled beer cans littered the dock; through the filmy water I could see more trash lying on the bottom, nestled in the murk. Turner grabbed my elbow and steered me away.

We walked down the dock and across the property to a raised platform that overlooked the sea pen. Climbing the flimsy stairs nervously, I thought I heard the bang of gunshots, but it was just kids popping balloons. From this perch we could see directly into the dolphin enclosure. While Turner went to get us drinks, I counted the bottlenoses floating listlessly in the water. There were seven.

Not a single dolphin moved; they hung vertically and motionless, with only their beaks and the tops of their heads visible. I sat down and took in the scene. By this point in my travels I had encountered a lot of dolphins. I had seen them in their element and I had seen them in captivity, but I had never seen dolphins in such a dire state. It was a miserable spectacle of cruelty, the net result of a cascade of grasping interests. To anyone with eyes, it was obvious these bottlenoses were dying.

Turner came back with two SolBrews and sat down. His face looked strained. "I'm gonna swim in there one night and cut the nets," he said, in a low voice. We watched numbly, "La Bamba" braying in the background. Below, a gang of kids leaned against the chicken wire fence. Every so often, one of them would pick up a stone and throw it into the enclosure. The dolphins continued to hang at the surface, bunched together and staying as far as possible from the people.

Evening proceeded and the falling light cast moody shadows across the water. A young man in a black shirt that said SECURITY squatted on the dock holding a pail. It must have been feeding time, but the dolphins made no moves toward him. We watched as he tried to coax them over, without success. None of the bottlenoses even turned their heads; they huddled against the wire, facing the open water. It was the first time I had ever seen dolphins turn down fish. His only taker was a vigilant pelican.

Turner and I stayed until the dusk was gone, not saying much, just watching the dolphins floating inertly in their pen. After a while, a band started up on the dock, the din of electric guitars yowling into the night. The kava drinkers were still there, slumped at their tables. One man lay on the floor, out cold, with his pants partway off. As Turner started up the dinghy, I took a last look at the seven captives. Their plight was heartbreaking, infuriating—but I knew that tomorrow might hold even more distressing visions. In the morning, Makili and I were going to Fanalei.

≈

"Okay, so we're checking out the coastline for tsunami damage. That will be our story." Our pilot, Glenn Hamilton, was making his final check of the helicopter before we lifted off. He seemed nervous, which wasn't comforting. Makili and I sat in the four-seater bird wearing headsets and bulky orange lifejackets; him in the copilot's seat, me in the back. The sky was a low gray ceiling, tinged acid-yellow by the sickly light of garbage fires.

At the airport I'd told Hamilton, an Australian who resembled Tom Cruise, the details about what we hoped to achieve; that we wanted to swoop down over Fanalei, getting a good look at the village and checking for signs of dolphin hunting. We would circle a bit, enough to document the scene—then we would scram. Hamilton had not taken this plan well. "We have a lot of trouble with Malaita," he said, frowning.

"About what?" I asked.

"About everything."

In particular, he balked at the idea of repeated tight passes over any villages. Whenever helicopters surveyed the island too intently, he told us, village chiefs bombarded the pilots with demands for compensation. Anyone traversing their airspace without payment, they claimed, was trespassing.

Makili explained the dolphin situation—at least 1,500 animals gone during the past two weeks—and eventually Hamilton settled into the idea of flying over the area, using a recent tsunami as a cover. He was shaken by the dolphin news: "I heard something about that on the radio," he said, wiping his brow. "But I didn't believe it. I mean, to kill a thousand dolphins in one day seems impossible."

"Nah," Makili said, shaking his head. "They can do more than that."

It was O'Barry who had suggested the flyover, describing the alternative—an overnight boat ride—as a never-ending ordeal. Having already endured a four-hour cruise in these waters, I wasn't eager to repeat the experience at length. Though air travel around the Solomons is patchy in general, it hadn't been hard to find a helicopter company here, because of the mining and logging interests. Booking Hamilton, I'd felt, was the right approach. I didn't want to hang out with expert dolphin-killers, much less listen to their grievances. To date, I'd managed to avoid malaria and dengue fever and I preferred to keep it that way. I was completely fine with buzzing Malaita rather than hacking across its hinterlands. One backpackers' guide had cautioned even the most intrepid adventure seekers to think twice about that island, referring to it as "the world's end." I had also read that in some Malaitan settlements it was a grave insult for women to show up wearing clothes: females had to visit naked. This didn't sound overly appealing.

We spun up into a horizonless haze, winging south over Guadalcanal. Endless tracts of palm-oil plantation unfurled below us, a man-made monoculture fast replacing the native forests. Every acre and every tree was identical, as though stamped from the same factory mold, with miles of straight roads slashed between them. Now and then we passed over gaggles of rusted-out tanks. "This sort of stuff's all over the Solomons," Hamilton said, through the headset. A burst of radio activity caught his attention, and as he listened he suddenly turned the helicopter, veering left: "There's a bit of chatter on the radio in Honiara," he told us. "There've been some bomb blasts. I just have to get us around that area."

"*What* bomb blasts?" I said.

Hamilton shrugged. "They find bombs. There are quite a lot of them. They just blow them up."

He adjusted the heading and soon we were flying over Ironbottom Sound. Malaita lay before us, snaky and forbidding. Inland rivers carved their way to the ocean. The scenery was dramatic, but I was preoccupied by the fact that my door wasn't closed properly. At my feet, the water below could be seen through a half-inch crack. And something was rattling—loudly. "Sorry about the unpleasant noise," Hamilton said. "Nothing we can do about that."

Makili sat silently, staring out his window. He didn't seem to notice the anvil clouds hunkered above Malaita, so I worried enough for both of us. Hamilton was busy, fiddling with instruments. Rain began to pelt the windshield and he glanced up, and turned course again. We skirted the edge of the thunderheads, flying above the mist that streamed off the mountaintops. The ocean changed colors like a mood ring. We rattled on.

A half hour later, Hamilton gestured toward a club-shaped peninsula ahead. "That's Fanalei," he said. "I'm going to fly wide of it, and then I'll pass above it twice. I'll go as low as I can." Makili turned to me in the backseat. "There's the lagoon where they drive the dolphins in to slaughter," he said, pointing down. It was a setup reminiscent of Taiji, the coastline forming a ready-made dolphin pen. Once the animals were chased in there, they'd be trapped. I could see dugout canoes heeling in the shallows, and a dark oily slick coating the lagoon's surface. Near shore the water was brown, rather than blue.

I was focused on the village, trying to make out detail, when I saw something flash from the trees. Then, nearby, another light flared, and another. Hamilton flinched. "Ohhhhh, this is what I was hoping to avoid," he said, explaining that the tribesmen were shining mirrors at the sun to signal their displeasure. "What does that mean?" I asked. "Well," Hamilton said, "it means this is not a good place."

On the next pass we dipped lower. I got a closer look at Fanalei's straw-colored, thatched houses. The construction was primitive. Everything was built at sea level; even moderate waves could wipe away the settlement. Below, a crowd had gathered, waving at the helicopter with agitation. Swaths of debris were visible on the mud flats, though whether they were the remains of dolphins I couldn't say. Supposedly the villagers ate the meat, but what did they do with the viscera? I suspected they threw the skin and entrails back in the water—there was nowhere to bury anything. I knew from the Al Jazeera footage that Fanalei was knee-deep in dolphin bones: it was impossible to walk anywhere without tripping over a spine someone had tossed, or feeling the crunch of beaks beneath your feet. Makili surveyed the village. "If they had caught any today they'd be cutting them on the beach," he said.

"We have to go now," Hamilton said, with urgency. "I'm going to hear about this." He turned inland, and away from Fanalei. I looked back at the serpentine coastline, and the barely unsubmerged land. The last thing I saw before the village faded from view was a tribesman in a dugout canoe. The man paddled furiously, racing into the lagoon as if to follow us. Then he dropped the paddle and stood up, shaking his fists at the sky.

≈≈≈

"Ah, these dolphins are sick. They're really sick." Makili examined the Kokonut Café photos I'd shot, scrolling through them on my camera. He looked up. "They saw you taking these?"

"Maybe they did," I said. "But they were facedown by five o'clock. We got away clean."

"Yeah," Turner agreed. "That kava must be pretty strong."

We sat on Turner's boat in the early evening, drinking yet more SolBrews. It was my last night here, so Turner had invited Makili and me over for dinner. We lounged on the bow, joined by Turner's Portuguese water dog, Sal.

I wasn't the only one leaving. Turner planned to weigh anchor in a few days, and Makili had a flight in the morning to Ghizo, an island in the Solomons' western province. His fiancée awaited him there, and he had unfinished dolphin business in the area as well. He had referred to this before—and I knew there was something personal in the account—but now I asked him for the whole story. "Ah, nobody cares about it," Makili said, shaking his head.

"Yes we do," Turner said. "Do you need another beer?"

Makili chuckled. "One thing I want to tell you. Don't ask a Solomon Islander if they want a beer. Just give one. They'll drink."

Leaning back against a red buoy, Makili began his tale. Not long before I came to Honiara, he had learned that a village on Kolombangara Island, Ghizo's next-door neighbor, had captured a pod of bottlenoses. "They have a cove," he explained. "It's much bigger than Gavutu. The dolphins swim in naturally, that's what they do. Now the villagers have realized that dolphins are worth a lot of money. This dolphin trade business—they've heard about it. So when this pod came in they closed off the entrance. And they held them there for almost a month, captive. When I heard about that, I straightaway went down there."

Makili arrived at Kolombangara and found fourteen dolphins, only half alive. Others had already died. The villagers hadn't fed them: they'd fully expected someone to show up promptly with wads of cash and whisk the dolphins away to the airport. "I spent a week negotiating with the community," Makili said. He was threatened and menaced; the people were not willing to let the dolphins go without a payout. As Makili described that time, his shoulders began to tremble. Each day, he would go to the lagoon and see the dolphins languishing and feel powerless to help them. "The people denied the very fact that they were suffering," he said, his voice ragged.

Turner and I were quiet, listening. Slow reggae music played in the background. Makili crossed his arms, as though hugging himself, and continued. "I walked down to the water," he said, "and this little calf came straight to me. He lay on my hands. I could see a lot of bruises on him. You could smell it . . . his body was infected." He paused to collect himself, but tears had begun to roll down his cheeks. "I spent eighteen minutes in the water with him. I was crying. I was crying in the water and I was shouting." Makili looked at us; tears quivered in his mustache and beard, spilling onto his shirt. "The little one came with a message. He told me that I have to do something. *Do something* so that my family doesn't suffer as I am suffering. He came to me for his last breaths. Then he died in my arms."

Makili had walked out of the ocean holding the dead calf, and, with the full force of his emotions, demanded that the villagers release the rest of the dolphins. The next day, they did. He took the calf, whom he had named Little Jacob, back to Ghizo and froze his body. On this next trip, he would bury the dolphin in a gravesite a local landowner had donated. "Now I tend to believe that yes, dolphins do have feelings," he said. "They knew that I was trying to fight for their lives. They *knew* it."

Now we were all crying. I grabbed Sal and buried my face in her fur. The clouds were tinted peach and the air felt warm and soft, but the story was hard. So was the reality of the Kokonut Café, Gavutu, Fanalei—all the turmoil that had poured out, like poisonous smoke, from under the lid that had been pried open in this place.

I had a question, rhetorical perhaps, but it had been on my mind for days. "How do you think humans got so cruel?" I asked Makili. He gazed at the ocean, then back at Turner and me. "We forgot," he said, letting the words linger. "We forgot our responsibility. And we forgot that we are as equal as any living thing within the chain. There's no hierarchy in this. *Nah.* We are part of the same family: living things. All the rest of it is just totally fucking bullshit."

A sly quarter moon had crept into the sky while the daylight dissolved in a blaze of gold. I had planned to bolt for the airport tomorrow morning, but now part of me wanted to change my ticket and follow Makili to Ghizo for Little Jacob's burial. Or sail to Savo with Turner to find the elusive dolphin cave. It was unlikely I'd ever return to this country, but I realized that my reporting about it wasn't complete. A piece of the story was missing. Now, after seeing the wreckage he'd left behind, I knew that I needed to talk to Chris Porter.

CHAPTER 9

GREETINGS FROM HAWAII: WE'RE HAVING A BLAST!

Only two miles off the Kona coast, the ocean plunges a mile deep. The drop-off is a long, steep slope to the seafloor, a downhill lava ramp. There are few places in the world where someone can be standing at the marina and, before he's finished his coffee, find himself suspended over the abyss—Kona is one of them. Which is why Robin Baird, a biologist whose work is focused on deep-water dolphin species, spends so much of his time there.

While the spinner dolphins venture into the depths at night and then return to the shallows by day, other dolphins hardly ever approach the shore. As with all creatures who dwell in habitats inhospitable to humans, we know very little about them. They are exotic enigmas, as inaccessible to us as snow leopards or harpy eagles. About four times a year, Baird and his team cruise the blue water off the Big Island and Kauai, searching for these uncommon dolphins—false killer whales, pygmy killer whales, melon-headed whales, rough-toothed dolphins, Risso's dolphins, and striped dolphins—along with their more familiar brethren: pilot whales, bottlenoses, spinners, and spotted dolphins. Every so often he also encounters the dolphins' close relatives, beaked whales and sperm whales, and he gathers information about them, too. In all, Baird studies eighteen types of toothed whales that live in Hawaii.

I first met Baird on Maui, where he had stopped by to give a talk at the island's annual Whale Tales festival, a celebration of the thousands of humpbacks that migrate annually to mate and calve there. (During the winter months, these forty-ton creatures are so visible from any shore that it's hard to look out and *not* see them spouting and breaching.) Baird, a jovial fifty-year-old with light blue eyes, ginger hair, and the kind of fair skin that scorches easily, had just finished a two-week research cruise off Kauai. On the day we met for lunch, he was rosy with sunburn.

I wanted to learn more about the dolphin societies in the open ocean, and by every account, Baird was the man to ask. He is a prolific and well-known scientist who collaborates widely and can reel off details about studies, papers, and cutting-edge science news, reciting from memory as though he were reading from a teleprompter. Baird has an easygoing nature, but there is a jet engine running between his ears. "I've always been interested in rare species," he told me, describing an upbringing on British Columbia's Vancouver Island that was long on nature and wonder. From childhood, Baird was exposed to orcas, not just the captives who had lived at Victoria's Sealand of the Pacific, but the magnificent Northwest pods that swam near his home.

To anyone who has ever seen them in the wild, orcas leave a spectral impression, trailing far more questions than answers in their wake. For his PhD, Baird studied their hunting methods, strategies that require the animals to communicate precisely, cooperate extensively, and work to achieve a shared goal. Any long-lived species that can do these things, Baird said, should be counted among the smartest on earth. Knowing the scope of the orcas' sophistication, he was doubly fascinated when he saw his first false killer whale, or pseudorca, an even more cryptic ocean citizen. These whales are about two-thirds the size of orcas, with similar thick, conical teeth, a more tapered physique, and dark, muscular bodies. Descriptions of the species often contain adjectives like *fast, agile, emotional,* and *extremely clever.*

This wasn't just Baird's first glimpse of a false killer whale: it was the first sighting reported in all of Canada, for anyone, ever. The animal had stranded and didn't survive, but finding one on its own must have been a relief—pseudorcas form such close bonds that mass strandings are distressingly typical. Once, in 1946, more than eight hundred false killer whales died on a beach in Argentina; groups of more than a hundred have stranded in South Africa, Europe, Australia, New Zealand, Sri Lanka, and Florida, among other places. Baird's interest was stoked. He and his colleagues necropsied the body, extracting every bit of information they could from it.

The scientists found that the dead whale's liver was overloaded with mercury and DDT, off-the-charts levels of toxins. In the twenty-eight years since then, Baird has collected biopsies from forty other false killers and discovered a disturbing pattern: they're *all* highly contaminated. They are saturated with PCBs, dioxins, flame retardants, heavy metals, pesticides—some of the most noxious carcinogens ever unleashed. Many of these long-lived chemicals, known as persistent organic pollutants (POPs), have been outlawed for decades, but they still linger in the environment. "Over time they've been incorporated into food webs and dispersed across ocean basins," Baird said. "It's insidious. You can't see these chemicals and they're not killing animals outright, they're just making them more susceptible to infections, and if they get an infection they'll have greater difficulty dealing with it." Recently, Baird had helped lead the effort to get false killer whales included on the endangered species list, proving with painstaking science that their numbers are fast declining.

False killer whales also suffer from a problem that plagues dolphins all over the world, from tiny Hector's dolphins right up to orcas: as apex predators, their diet consists of the same marine creatures we like to eat, and in recent decades we have hoovered the oceans, wiping out one fish population after the next with our destructive methods. Overfishing, bottom trawling, taking fish before they're old enough to reproduce, netting and longlining practices that kill indiscriminately—these idiocies have landed both humans and dolphins in a pinch: we're running out of seafood.

Though dolphins have been hunting in the planet's waters for eons longer than we have, they are no match for our onslaught. But they want tuna and salmon and squid too. Where dolphins and commercial fisheries meet, the results can be devastating for dolphins, not just in Japan, where the surreal notion that cetaceans are gluttonously taking all the fish—as opposed to industrial fleets the size of cities scouring the seas with radar, sonar, fish aggregating devices, and spotter aircraft—has cost countless dolphin and whale lives; but even in Hawaii, where most of the false killer whales Baird studies, in three roving pods, have visible injuries from fisheries. Their dorsal fins are lacerated and deformed, sometimes all but razored off, from getting entangled in longlines. Recently, Baird told me, a false killer whale had washed up dead on the Big Island with five fishhooks in its stomach.

≋

Taken to their extreme, fishermen's interactions with dolphins can become vengeful. In places where fishing quotas have been lowered due to the collapse of stocks, the animals have even been the targets of hate crimes. In 2012, for instance, dolphins began washing ashore in the Gulf of Mexico, their bodies mutilated. One bottlenose had its tail cut off; another had its jaw severed. Several other dolphins had been shot or slashed open, and one featured a screwdriver stabbed into its head. After seven of these carcasses were found, people began to speculate that a crazed dolphin serial killer was on the loose, with a $30,000 reward offered for information. RACE ON TO FIND GULF COAST DOLPHIN KILLERS, reported CNN. WHAT'S BEHIND SPIKE IN GULF COAST DOLPHIN ATTACKS? asked *National Geographic*. But the bodies were sprinkled across the region, from Florida to Louisiana, and scientists believed that, far from being the work of a single madman, the cause was more widespread: frustrated fishermen.

Similar dolphin-loathing sentiments fester in other places. Down in the Amazon, scientists have become alarmed by the rapid disappearance of the region's botos, or pink dolphins. For ages, the animals were respected and revered by the local people; now, botos are hunted and used as bait to catch piracatinga, a tasty catfish, and reviled for damaging fishing nets. To find out if the botos were being deliberately targeted, researchers interviewed sixteen fishermen, letting them speak anonymously. Eleven of the fishermen admitted to killing pink dolphins at every opportunity. "Nobody likes the botos," one man reported. "They are river pests that need to be exterminated. They only cause harm." "If we let them reproduce," another explained, "we will not have anything left for the humans."

Even in the Hawaiian waters Baird surveyed, which seem at a glance to be sparkling clean and blue, there were vexing signs that neither fishermen nor dolphins were thriving. Tuna populations have plummeted, while the number of longline hooks has quadrupled since 1996. (The boats set miles and miles of these hooks, and then haul in anything that happens to get caught on them, including dolphins, sharks, seals, turtles, and even birds, discarding as much as 40 percent of their catch.) Other fisheries are in worse shape. In the Gulf of Mexico, which supplies much of America's seafood, conditions have verged on apocalyptic. Even before BP's Deepwater Horizon well blew in April 2010, gushing an estimated 150 million gallons of crude into one of the most fertile ocean nurseries on earth, the Gulf had been

under siege. Pollution, drilling, agricultural runoff, wetland destruction, bottom trawling, overfishing—the area has taken hit after hit.
The Gulf is also home to the Atlantic's biggest dead zone, a nearly
8,000-square-mile area devoid of oxygen. None of this is good for the
catch, which has been dwindling rapidly for the past fifteen years. But
the BP disaster, in its biblical scope, pushed ecosystems into uncharted
territory. Murdered dolphins, while grisly and sad, were not the worst
fallout scientists were observing.

In the years since the epic spill occurred—and the toxic, volatile
(and little studied) dispersant, Corexit, was sprayed with abandon in
its aftermath—the tab for our negligence has come due, and it is being
picked up by marine life. Despite public relations' assurances that
everything is just fine, that the oil has miraculously disappeared, eaten
by microbes perhaps, but in any case long gone, evidence has mounted
that dolphins, along with fish, shellfish, corals, and even plankton,
have been severely contaminated. Along the Gulf Coast, shrimp and
fish without eyes, or even eye sockets, have been hauled in, gaping
lesions are commonplace, crabs' shells are not developing properly,
corals are smothered in brown gunk, and dead dolphins have rolled up
in appalling numbers from Texas to Virginia.

One startling effect was a flood of babies and fetuses, more than
eighty stillborn bottlenose calves coming ashore in the Gulf during the
first four months of 2011. There was an influx of dead adult dolphins
too, beginning even earlier, three times the number found in an average year. A cold snap before the spill didn't help the dolphins' defenses,
but it defies logic to think that swimming in a stew of contaminants
wasn't a major cause of the spike. For the past five years (and counting)
the National Oceanic and Atmospheric Agency (NOAA) has been
monitoring an "Unusual Mortality Event" among the Gulf's dolphins.
Though it is only one of sixty such die-offs in the United States since
1991, it is a doozy. Between May 2010 and May 2015, 1,199 dolphins
have washed up dead. Those are only the ones that we've found. Given

that most dead dolphins don't ever make it to shore, their bodies sinking in the deep or being eaten by predators, scientists estimate that the real number of dolphin casualties could be fifty times higher. And the bodies keep coming.

In Barataria Bay, Louisiana, a formerly rich estuary inhabited by bottlenose pods, a 2013 study of thirty-two dolphins found that most of the animals were grievously ill or dying. They had lung and liver disease and depleted hormones; they were underweight and anemic. Many of the dolphins had lost their teeth. "I've never seen such a high prevalence of very sick animals," said NOAA's Lori Schwacke, the lead scientist. "What we are seeing is consistent with oil exposure."

In the face of these facts it's hard not to get infuriated, to become overwhelmed. But one of the intriguing things about Baird is his measured outlook. While he feels strongly about protecting the animals he studies, he is never strident. Baird's philosophy when dealing with less dolphin-friendly interests is that it's better to collaborate with them, seeking to minimize harm to marine life (which can conceivably be done), rather than eliminate threats (which is probably impossible). "I tend to be fairly pragmatic and recognize that the world is a complicated place," he told me.

Recently, Baird had been part of a "take reduction" team put in place by the National Marine Fisheries Service (NMFS) to investigate ways in which fewer false killer whales and other dolphins might get snared in Hawaii's commercial fisheries. Included on the team were a number of fishermen. When I asked if a spirit of cooperation had prevailed, Baird hesitated. "Uh . . . yessss," he said, with a dry smile. "Not necessarily as much as you'd hope." But it was important to recognize the interests of a multibillion-dollar business, he stressed, if you wanted to get anything done. In the end, an agreement was reached in which the longliners will stay farther away from the main islands, and adopt circular hooks that are easier for the dolphins to dodge. "I eat fish," Baird said, "and I think fishing should continue. I just think it should be done sustainably and in a way that doesn't impact protected species." Perusing the menu, he recommended the mahi-mahi: "It's fast growing and has high reproductive potential. It only lives a couple of years so it has done fairly well in response to the overfishing of a lot of other things, like tuna and swordfish." A short life span also means the mahi-mahi will spend less time marinating in ocean pollutants.

Baird's collegial spirit also extends to the captivity industry, though like other scientists he is aghast at Japan's dolphin hunts. "I'm not one of these people who think you should release all the captive dolphins, but I definitely think there should be no captures from the wild," he said, describing how, as a kid, aquarium and zoo visits had sparked his fascination with animals. "I guess if I were in charge, I would force them all to have a really strong educational and conservation message. For some species, like bottlenose dolphins, I'd be okay with it as long as they were looked after properly. But I don't think they are."

When trying to find a reasonable balance in dolphin-human affairs, Baird's most challenging constituent is also one of his main sources of funding: the U.S. Navy. On one hand, the Navy, America's steward of the seas, has bankrolled more ocean research than any other entity. On the other hand, it does so for its own ends, or because it is forced to by the courts. Many of Baird's investigations in Hawaii, for instance, are fact-finding missions to assess the toll of the Navy's increasing use of mid- and low-frequency sonar around creatures whose lives depend on their acoustic senses. Sonar is the Navy's main tool, used to detect the presence of enemy submarines or other underwater threats, but it is also one of the loudest noises in the sea. The brawniest sonar systems can flood the water with 236 decibels of sound, about the same head-splitting intensity as a rocket launch.

The quest for more information about why—in the vicinity of undersea war games—dolphins and whales have often stranded en masse with blood streaming out of their ears and eyes, was not rooted in the Navy's concern for the animals. It was spurred by lawsuits from

environmental groups, not to mention public fury. Baird's research—
and studies by other scientists—has pinpointed areas where the most
vulnerable cetaceans live, so the warships could ideally steer clear of
them. Perhaps the Navy could build its $127 million Undersea War-
fare Training Range somewhere other than the only known calving grounds of critically endangered North Atlantic right whales, or
refrain from missile tests within a mile of American shores. Maybe we
could set aside small refuges? There were, after all, 139 million square
miles of ocean out there. For peacetime training exercises anyway, it
didn't seem like too much to ask.

Amazingly, it was. Though military sonar is proven to wreak havoc
on cetaceans—driving them away from feeding areas, causing them to
beach in panic, interfering with their communications and mating and
resting, bubbling up hemorrhages and embolisms in their organs—the
Navy has battled all the way to the Supreme Court to avoid even the
most modest restrictions. Its stance, unwavering and defiant, is that it
will strafe the oceans wherever and whenever it wants—and anyone
who disagrees with it is jeopardizing national and even global security.

Far from taking steps to reduce its impact, in 2012 the Navy submitted a five-year proposal to dramatically ratchet up its war exercises.
Plans included 500,000 hours of sonar operations and the detonation
of 260,000 explosives in Hawaii and Southern California alone. By
the Navy's own estimates, this would result in 155 dolphin and whale
deaths in these waters, along with 2,000 serious permanent injuries,
and some 9.6 million instances of hearing loss and disruptions of vital
behaviors—a 1,100 percent increase it referred to as "negligible."

≈≈≈

For years before the body count of brain-pulped cetaceans became
too high to sweep under the rug, the Navy denied its sonar had even the
slightest effect on dolphins and whales. As for underwater bomb deto-

nations, the noise was as benign to them, one Navy scientist claimed, as someone dropping a book onto a table would be to us. Given its record of understatement, the fact that the Navy's own harm predictions were so high meant the real damage to marine life was likely to be far greater.

After the NMFS rubber-stamped the Navy's proposal, a coalition of environmental groups sued it, and won. In a March 2015 ruling noted for its unusually strong language, the Navy and the NMFS were found to have basically ignored the Marine Mammal Protection Act, the Endangered Species Act, and the National Environmental Policy Act, and advised not to merely tweak existing plans, but to head right back to the drawing board. For anyone who cares if there are any creatures left swimming in the sea, it was a victory. But the matter was far from settled. In fact, the fighting had just begun.

These days, a major part of Baird's work involved placing satellite tags on the dolphins and whales who frequent the depths, where they are more at risk from sonar exposure. The tags track movements with precision. Baird can log onto his computer to see how the tagged animals react to Navy activities, which are widespread in Hawaii. Twice a year, for instance, off the northwest corner of Kauai, the Navy conducts its submarine commanders course, with days of constant sonar. And every other year, Hawaii hosts RIMPAC, the Rim of the Pacific Exercise, the world's largest gathering of international navies. During RIMPAC, the islands thrum with combat drills, mock battles, bomb blasts.

Our footprints were stamping through the ocean so fast that it was almost a race against time to gather data, particularly for the little-known species Baird studied. If there was any chance to save the fragile populations of false killer whales or melon-headed whales or beaked whales or numerous others, the time for intervention was at hand. Hard scientific information, while not necessarily adding up to power against Goliaths like the Navy, fisheries, and oil companies, was at least . . . something. It was a stone for David's tiny slingshot. It was a snippet of hope against a future in which only warfare and oil extraction and fish farming flourished in the seas.

Baird had already established the existence of resident melon-head whale, false killer whale, pygmy killer whale, and pilot whale populations in Hawaii; he had published papers about their ranges and their numbers, and had submitted them to the NMFS with his recommendations that their turf be spared from aggressive sonar use. He had tagged 218 dolphins and whales—and he had watched them adapt (or not) to the cacophony around them. "My feeling is that some species are really susceptible to impacts and others are probably pretty tolerant," he said, describing pods of bottlenose and rough-toothed dolphins that frequent the Navy's testing range. "Some individuals have probably been exposed to sonar their entire lives."

But the dolphins don't give up their secrets easily. Baird's work takes time and resources and, above all, patience. "One of my long-term interests is, how do you study animals that spend most of their time underwater?" he said, with an exaggerated shrug. The search for the nomads, the deep divers, the dolphins who might surface for a heartbeat and then vanish into the void, is not a task for landlubbers or the easily discouraged. On an average day, Baird starts at sunrise, ends in the late afternoon, and covers a hundred miles in a small boat.

The best way for me to understand his research, Baird said, would be to accompany his team on a field project. In three months he would be returning to Hawaii, coincidentally during RIMPAC 2014. This year the war games were slated to be bigger than ever: twenty-two nations would be participating, along with forty-nine warships, six submarines, more than two hundred aircraft, and 25,000 personnel. From late June through early August there would be torpedo firings, ship sinkings, cat-and-mouse submarine hunts—a sonar jamboree.

≈

"If I didn't warn you, we spend a lot of time just cruising around."

Baird, clad head-to-toe in sun-protective clothing, sat on the flying bridge of his twenty-seven-foot Boston whaler, piloting us toward the southern tip of the Big Island. We were a few miles offshore; the calm conditions were perfect for sighting the little disturbances that might be a dolphin's dorsal fin; even the blip of a fish breaking the surface was visible from afar. From the start the day had been hot, the sky layered with wispy clouds that only seemed to intensify the sun's glare. There were six of us on board, each with specific tasks, drilled in as though we were operating our own miniature RIMPAC. While on land Baird had seemed mellow, on the water he morphed into a cheerful task-master: every minute we were out here, not a single degree of ocean went unobserved. Each person's eyes swept a 180-degree stretch, back and forth and back and forth like windshield wipers. On the horizon I could see the dark outlines of naval destroyers and cruisers, hulking vessels that looked like something from a dystopian Xbox game.

Daniel Webster, Baird's right-hand colleague, stood at the bow. Webster was athletic and nimble, a midsize forty-year-old with brown hair tucked under a baseball cap. He was quiet but quick to smile, and he exuded the competence I'd come to associate with scientists who spend a lot of time on boats. There is a high degree of difficulty to performing intricate tasks on a tiny, rocking, heaving platform, and keeping sensitive electronic equipment functioning in corrosive heat and saltwater. When a Cuvier's beaked whale poked its head above water for maybe a second and a half, Webster was the one charged with tagging it, while Baird positioned the boat. This is about as simple as hitting an archery bull's-eye while balancing on a BOSU ball, on a conveyor belt.

Like Baird, Webster was an experienced birder. It was one of the most valuable skills a deepwater dolphin researcher could have, the line item on any biologist's résumé that was most likely to earn him a spot on Baird's team. Catching a distant glimpse of a hurtling object the size of a baseball and being able to discern minute details about the stripes on its wingtips or the notch on its beak was exactly the kind of laser vision required to spot, for instance, a far-off ripple that was actually a false killer whale coming up for a breath.

Facing Webster, scouting the expanse of ocean behind him, was biologist Brenda Rone. Tall and lean, with the muscles of someone whose work was physically demanding, Rone specialized in difficult whale assignments. She worked often in the cold, unruly waters off Alaska, the Arctic, and Canada's eastern coast, searching for blue whales and right whales, bowheads and humpbacks, maneuvering to tag their rolling bodies from the bouncy-castle flimsiness of dinghies. Today, she leaned against the whaler's windshield, eyes hidden behind polarized sunglasses, hair scraped back in a ponytail, and a gaiter pulled over her face to shield her from the remorseless sun. Rounding out the crew was Jim Ward, a high-spirited photographer in his twenties, and Kelly Beach, a sweet-natured, whip-smart intern from Baird's group, the Cascadia Research Collective in Olympia, Washington.

Photography was a critical part of the process. Stacked behind Rone were hard-sided Pelican cases, containing Canon 7D cameras with zoom lenses. When they occurred, the encounters with rare dolphins were fast and furious, and the most important thing, aside from tagging, was to document the animals who had showed themselves. For all three false killer whale pods he had met in these waters, Baird knew their members as individuals. Later, back on land, he would pore over the pictures and ply information from them: who was still around, who was missing, who had a new calf or a fresh dorsal fin injury, who was looking especially skinny. When there were so few dolphins left in a species, every observation mattered.

To raise the odds of finding his reclusive study subjects, Baird had created fliers he left with sportfishing boats, tour operators, scuba clubs—anyone who might happen upon a pod of false killer whales on the water. On the back, his contact information was highlighted, including his cell phone number. He also passed out laminated field guides with photos of the pseudorca's dorsal fins, each bearing its signature deformation. The guides were beautifully designed and crammed with facts about the species. Outreach is key, Baird explained, when you hope to raise concern for creatures most people have never heard of, and will probably never see. To draw children's attention, Baird had even made press-on pseudorca tattoos.

Because it was my first day, I had been allotted the seat next to Baird on the bridge. Up there it was partially shaded, a major bonus. My main task, not very demanding, was to hit a clicker whenever anyone spotted a wedge-tailed shearwater or a Bulwer's petrel, two seabird species. When a less likely bird whizzed overhead, Baird yelled out, "Photo!" calling the species on the fly: "Black-winged petrel!" Webster hung his body over the rail, his camera whirring on motor drive. While cruising, Baird noted everything, every sighting and perturbation and augury. In the past, the team had stopped to pick up bobbing trash bags and discarded fishing nets and other plastic debris, but after an average day the boat had become so laden with crap that the marina now forbid them from using its dumpsters.

The most common floating objects were balloons. Released during festivities, helium wafting them over the sea, these shiny Mylar orbs were killers. Turtles, dolphins, and other marine animals ate them, mistaking them for fish or squid, with fatal results. I had noticed a sticker on Webster's tagging case that read: BALLOONS BLOW, DON'T LET THEM GO! "Most of the balloons we pick up have some sort of party writing on them," Baird noted. When people fling their birthday wishes or their graduation congratulations or their wedding celebrations to the sky, they have no idea how lethal their actions are.

We motored slowly south, to avoid the trade winds. The light turned silvery yellow, portentous clouds massing above the volcano peaks. Webster came up and sat with us in the shade, because there wasn't much happening on the water. We stopped to net a dead squid, its deep maroon body bitten in half, and we saw frigate birds gliding overhead, a red-tailed tropic bird, and some ahis hopping about, but there was a distinct lack of dolphins. Several hours in, I was getting it: this was needle-in-a-haystack work. "We see pseudorcas about once every three weeks," Webster said. Baird nodded: "That's why, historically, there hasn't been a lot of research on the species. But we try to be productive the rest of the time. We'll work with whatever we find out here."

I had wondered what it was about false killer whales that so captivated Baird, and as we were driving, I asked him. Baird brightened; there were few things he liked to talk about more. Prior to his arrival in Hawaii, he and his wife, Annie, had driven from Olympia up to the Vancouver Aquarium, where a pseudorca calf was being rehabilitated after being rescued from the beach. In a fortunate reversal of Baird's first false killer whale stranding, the calf, who'd been named Chester, had survived, though he was only about six weeks old, still toothless and dependent on his mother's milk. Initially, no one thought Chester would make it, but so far he had, and now the aquarium staff and the public—anyone who'd met him—were equally smitten. The thing about false killer whales that surprises people, Baird told me, is that they are fascinated by us, "so much more than any other animal I've ever interacted with." He grinned affectionately. "They recognize humans as being something similar to them."

Two colleagues of Baird's, Dan McSweeney and Deron Verbeck, had spent time in the water with false killers, both men returning with extraordinary stories of how engaged the animals were. One time, McSweeney was diving near a pod when one of the pseudorcas swam up to him with a tuna in its mouth and spat it out at him, as though delivering a gift. He took it; the whale swam away and then boomeranged back. McSweeney handed the fish to the whale, who once again took it gently in his mouth. "It's not the first time this kind of thing has been documented," Baird pointed out. "It's basically ritualistic prey sharing. I say ritualistic because I think it has a symbolic component to it."

While other top predators such as sharks and lions often shared the spoils of the hunt, jointly ripping into whatever they'd caught, Baird had observed false killer whales simply passing a fish around among themselves, without anyone taking the slightest nibble. "I don't know of any other species that does this," Baird said. "So, why would they? Well, they're hunting companions. They've probably been hunting together for years. They're buddies. It's a way of demonstrating trust. And the fact that they give it back . . . it's not like they're trying to get more of the fish themselves. The purpose of this isn't just to eat the prey. There's something more: it's part of their culture. And you know, what's going on in the head of an animal that actually brings food to another species?" He smiled. "I have a dog. He's really cute. But when I give him a bone, he's certainly not going to offer it to me."

By now we had cruised thirty-five miles to the southern tip of the leeward coast, and Baird began to turn the boat around. We'd been trolling through water four thousand meters deep, and now we would make our way back at a slightly shallower depth. So far, no dolphins. I could see where the lava from Mauna Loa had rivered to the sea, with an isolated fishing village, Miloli'i, perched at the shoreline. Baird had just opened a package of Ritz crackers when his phone rang. "We're in the southern part of Kona," I heard him say. And then I watched his face turn to stone.

The call was brief. Baird hung up and turned to Webster and me. "As we speak, the Navy is using mid-frequency sonar," he said, in a tight voice. The caller, Baird told us, was a dive captain who'd heard the sonar reverberating through the hull of his boat: "Divers are getting out of the water because it's too loud and painful." Worse, this was all happening near Puako, a patch of ocean Baird had repeatedly urged be categorized as an exclusion zone. It was home to a resident pod of

450 melon-headed whales, a tribe that did not tolerate war games well. After sonar lit up the depths during the 2004 RIMPAC, two hundred melon-headed whales stranded in Kauai's Hanalei Bay.

Baird, agitated, began to make more calls. He was visibly upset. "In my opinion, it's really irresponsible of the Navy," he fumed. "It's a PR mistake, big time. They could go somewhere else—and they go where there's a sensitive species and people have to get out of the water?" Baird stared at the ships in the distance. His face was flushed, and not just from the heat. "How can that be good for *anyone?*"

≋

The discovery that certain animals navigate their worlds by relying on sound rather than vision dates back to 1773, when an Italian scientist named Lazzaro Spallanzani noticed that bats could fly expertly in total darkness, swerving around obstacles they couldn't possibly see. Astonished by this, and somewhat spooked as well, Spallanzani caught some bats and began to experiment with them.

You would not have wanted to be one of Spallanzani's bats. Though his first experiment was sort of cute—he outfitted the creatures in tiny hoods—Spallanzani quickly moved on to heavier-handed efforts, blinding them and even removing their eyes, coating their wings in varnish and flour paste. But once the bats recovered from their injuries, they could still hunt with ease, even eyeless. Spallanzani was stumped. He moved on to their ears, plugging their auditory canals with hot wax. This time, success: the bats bumped into things. Without their hearing, it seemed, they were lost. When Spallanzani presented his findings, surmising that bats were sonic savants without much use for their other senses, he was laughed at by some of the most renowned scientists of his day. His theory was shelved for a century, referred to dismissively as "Spallanzani's Bat Problem."

But Spallanzani was right. Though it wasn't until 1938 that his

"problem" was solved by scientists at Harvard, even then the phenom-
enon of animals that could echolocate, or produce their own biological
sonar, was poorly understood. In 1947, Arthur McBride, the curator of
Marine Studios, pointed out that whatever this extrasensory skill was,
dolphins appeared to have it too. The details of how this all worked
was still anybody's guess. It was only in the sixties—when Ken Norris
presented the results of experiments in which he'd blindfolded dolphins
with rubber suction cups and watched them move effortlessly through
an underwater maze to locate a scrap of fish—that dolphin sonar was
proven to exist. By this point, of course, dolphins had been echolocat-
ing for about 35 million years.

Basically, active sonar involves sending out a beam of sound
and then analyzing the returning echoes to determine the three-
dimensional physical properties of an object in space, or underwater.
Different materials reflect sound at different wavelengths; some items
are bouncier and sonically brighter than others. Others absorb sound,
sending back weaker echoes. All sonar systems, regardless of finesse,
consist of three parts: a transmitter, a receiver, and a way to process
the signals.

Scientists soon learned that dolphin sonar was anything but basic.
As one researcher put it: "To say that dolphins echolocate is like saying Michelangelo painted church ceilings." The animals generate
ultrasonic clicks using structures in their nasal passages (near the
blowhole); a fat-filled sac in their foreheads called a melon focuses the
sound. When the clicks hit an object the dolphins receive the echoes
through their lower jaws, which are also lined with fatty liquids. From
the jaws, the acoustic feedback is transmitted up to their ears and into
their brain, where it's interpreted and relayed to other senses, such as
vision. Their clicks emanate in a stream, up to two thousand clicks per
second, but dolphins can both aim and adjust each click individually,
changing direction, volume, and frequency—a feat of unimaginable
precision. They can even send out two click streams at once, in dif-

ferent directions, at varying frequencies. Using this sense, dolphins can detect minute variations in the size or composition of identical-looking objects, even at a distance. It's a spectacular system, ideal for life underwater, where light is scarce but sound travels easily, 4.3 times faster than it does in air. When manmade noises flood the dolphins' environment, it's like us being blinded by light so bright that we can't make out anything. But sounds are pressure waves—battering walls of energy—so for a more appropriate analogy, add to that our retinas painfully exploding in the glare.

Through the decades, countless experiments have been done to tease apart the inner workings of dolphin echolocation, particularly by the U.S. and Russian navies. There are obvious military advantages to getting an edge in sonar technology: since the fifties the Navy has conscripted dolphins and made strenuous efforts to harness their underwater superpowers. Even if we couldn't replicate their advanced sonar, we could certainly put it to use.

To date, bottlenose dolphins have served the United States in Vietnam, Iraq, Iran, Bahrain, Norway, Eastern Europe, and by guarding the Trident nuclear missile base in Bangor, Washington, to cite some known deployments. America has approximately seventy-five bottlenose troops; in the past this number has been higher. Their main tasks—according to the Navy, which declassified its marine mammal program in 1992 but is still less than chatty about it—have been to provide surveillance around ships, and to identify underwater mines so they could be cleared. Beluga whales and sea lions, deep divers accustomed to cold water, have also retrieved lost missiles from the seafloor. (At one point orcas were also trained for war, but their tendency to disobey orders made them less reliable soldiers.)

Officially, the Navy has always maintained that the dolphins, who are referred to as "Marine Mammal Systems," were there for nonlethal purposes, including swimming around with cameras clamped to their heads. But there were rumors of more sinister missions, recounted by disaffected trainers. In 1973, Michael Greenwood, a dolphin expert who claimed to have worked with the CIA in Key West, alleged to Morley Safer on *60 Minutes* that the animals had also been prepared for "swimmer nullification" work, shooting enemy divers with explosive darts and .45 caliber bullets. "The dolphins are being trained so that they can do any variety of exotic tasks," Greenwood said, adding that many finned operatives had gone AWOL: "We have lost many animals. They run away very frequently." "The military's interest in dolphins is no Disneyland scenario," Safer concluded somberly.

Regardless of what assignments the dolphin troops have been given, now or in the past or in the future, the Navy has maintained that its dolphins "will be provided the highest quality of humane care and treatment." On its Web site, in fact, it stresses that Congress has mandated this by law. No one should worry about these soldiers, the Navy says: "Marine mammals are actually in more danger from sharks, and . . . put in much more danger by people who feed them (which is why it is illegal)." Our Navy dolphins have taught us a lot, the Web site asserts earnestly, and "the more we know about marine mammals, the better we can protect them." All of which is very reassuring, if you are inclined to believe an institution that is busily blasting the animals out of existence, using the same sonar technologies that we recruited them to help us develop.

≈

The sky the next morning dawned pink and windless, and Baird motored out of the marina with purpose. We were going north to Puako, and we were going there with cameras and hydrophones ready. Even from afar, RIMPAC was easy to spot: a parking lot of gargantuan vessels. Before long we ran into a pod of bottlenose dolphins. Baird eased off so we could take photos; Webster dropped the hydrophone. It bobbed in the water, a fluorescent flag signaling its location. Knowing that sonar had been used nearby, Baird wanted to find out if its pings were still audible.

The bottlenoses swam up to the boat, game for a little wake riding: their bodies looked green below the water. I had been stationed on the bow, so I could see them closely. Many of the dolphins were marked with scrapes and cookie-cutter shark bites, round flesh wounds gouged out and healed over. One dolphin's dorsal fin had been sliced in half.

Standing on the bridge, Baird picked up his binoculars and scanned the horizon. "There's a Navy vessel at one o'clock," he said. "Second Navy vessel at two o'clock. That's either a cruiser or a destroyer. They would both have tactical sonar 53C," he explained, "but if they were using it right now, we would hear it through the hull. So I don't think they have it on." Another lummox of a ship was hunkered in front of Kohala's Four Seasons hotel: "That one looks like the back deck lowers into the water so they can run amphibious vehicles into it," Baird observed. As if to confirm his guess, a black hydrofoil emerged from the ship's belly and shot across the sea, sending up rooster tails of spray. Three oversize helicopters had also appeared, sweeping in loud and low, beating the ocean into a froth. If you were a military buff this parade might be exciting, I thought, but it wasn't the kind of scenery most vacationers longed to see from their $800-a-night ocean-front bungalows. As Baird had pointed out, it was a head-scratcher of a choice for a mock combat location.

We waited ten minutes longer, then Baird steered us back to the hydrophone so Webster could retrieve it. As we approached, we could see something floating beside it. "What?" Rone said. "It's an eel!" And it was: a snowflake moray, a reef-dweller that hides itself in crevices. Usually, the last thing an eel wants to do is leave its protective housing, but this one was snaking vulnerably across the surface, wrapping its gold and brown–striped body around the hydrophone as though desperate to escape the water. "What the hell is it doing up here?" Baird said. "This is bizarre."

When loud sounds boom through the ocean, all animals suffer, not just noise-sensitive cetaceans. Underwater detonations have a hellish history: in the past, the U.S. and Russia have even conducted nuclear tests in the ocean, generating aquatic mushroom clouds that exploded thousands of feet into the air, lighting the skies afire, boiling the sea, shooting out shockwaves and tsunamis that engulfed ships and even entire islands, miles away. You can read endless accounts of these whiz-bang experiments and never encounter a single reference to their toll on the living creatures below the surface, so their devastation can only be imagined.

Not all damaging underwater noises are percussive or ear-shattering: some, like commercial ship propellers, drilling, dredging, and cable-laying, are a constant, droning backdrop, low frequency vibrations that reverberate for miles. (Imagine living round-the-clock with a chugging air compressor strapped to your head.) Scientists refer to this ever-present racket as "acoustic smog," and they estimate that its levels have risen tenfold over the past twenty-five years. Industrial noise clangs and roars and hums through the oceans, obscuring the natural sounds animals use for mating, hunting, avoiding predators, navigating, migrating, communicating—really, for everything in their lives. As they try to escape the din or struggle to cope with it, they become chronically stressed and susceptible to illness.

Supertankers rumble by at about 180 decibels below the surface. At 185 decibels, human eardrums will burst; 200-decibel shock waves will blow a cow off its feet and cause death on land. But the loudest undersea noise of all, even beyond an atomic bomb detonating (Hiroshima was measured at 248 decibels), comes from the air guns used by oil and gas companies to prospect the seafloor. To conduct these surveys, dozens of cannons are towed behind ships, discharging 250-decibel blasts roughly every ten seconds for months at a time, affecting marine life over vast areas of ocean. These explosions can deafen creatures, cause organ failure, and harm larvae; they can silence whales at enormous distances and keep them from feeding, and cause nearby cetaceans to strand. At any given time, scores of these oil surveys are under way in every ocean of the world, and their numbers are rising.

As we wondered at the freaked-out eel and snapped pictures of it, the radio came alive, squawking with static. "*Securité, securité. This is Warship 47. Conducting speed limiting engineering casualty control drills. Request all vessels maintain safe distance.*"

Heeding their warning, Baird turned the boat and drove farther out to sea. Before long, we found another expired squid, this one speckled with white. As Webster bagged it, Baird watched the ships through binoculars. The sunlight glanced hard off the water, frying us. "Do you want some of my sunblock?" Ward asked me. "It's like superglue." I took some and coated my ears and face, again.

Webster, whose eagle-eyed vision rivaled Baird's, suddenly pointed in front of us. *"Kogia!"* he said. This meant nothing to me, but everyone else snapped to attention. A black back broke the surface, followed by a blunt head. Whatever it was, there were two or three of them. They disappeared quickly, but Baird had seen enough to confirm the sighting: dwarf sperm whales. "This is one of the species that's sensitive to Navy sonar," he said with a frown. "There was a stranding on Kauai during the submarine commanders course." We waited for almost an hour, but the whales didn't show themselves again. "It's an extremely poorly known species," Baird explained. "We've got 115 individuals in our catalog and it's the only long-term study in existence." The animals, he noted, were tough to track, skilled at shaking observers off their tails: "There's no relationship between the way they're traveling when they dive and where they come up. Often they'll go in a completely different direction."

≋

Before I left, we would find more galumphing pilot whales, a raucous bunch of rough-toothed dolphins, and, most thrillingly, a pod of three hundred melon-headed whales. Baird had gotten a tip from a sportfishing captain and we found the melon-heads twelve miles off shore. Once we arrived, they were impossible to miss. They were the most convivial dolphins I'd ever seen, inquisitive and eager to play with the boat, spy-hopping, streaking past us, and then logging at the surface in bunches. "They love to surf," Baird said. "When you accelerate even a little bit, you'll get twenty of them riding the bow."

Melon-heads live in expansive groups, and they appeared delighted with one another's company. They were everywhere, and Webster managed to tag two of them and collect biopsies. Later, Baird would inspect the results and share them with other scientists, including geneticists and toxicologists. This was not the resident pod he was worried about: these were transients, moving throughout the island chain. A band of rough-toothed dolphins had joined them, and weaving among all the dorsal fins were two sharks, an oceanic white-tip and a silky shark, their slinky bodies visible at the surface. The rough-toothed dolphins liked to jump, pogoing high in the air, and they had calico patches of pink on their skin, and long, narrow beaks. If the bottlenoses brought to mind powerful BMWs, the rough-tootheds were Porsches with front-end spoilers. The melon-heads were more like souped-up Mini Coopers.

The pods stayed with the boat for hours, everyone gunning their cameras, and Baird calling out directions from the bridge to help Webster aim his tagging gun, which resembled a miniature crossbow, as hundreds of dolphins darted around us and under us. (The last thing anyone wanted to see was a $5,000 satellite tag missing its mark and sinking into oblivion.) "Big male approaching you at two o'clock— target species," Baird shouted, and then chuckled. "We affectionately refer to certain individuals as 'targets.' Later, we hope to refer to them as 'tagged.'"

"They're flippy little bastards," Rone said, smiling. Webster nodded, wiping sweat from his forehead. "There'll be nothing, nothing, nothing," he said, "and then—mass chaos!" One melon-head swam alongside the boat, right where I was standing, and opened his mouth as though he were laughing. They were jet-black, happy-faced little whales, and I imagined that they were talking and joking and messing around, socializing as merrily as people at an excellent cocktail party.

≋

One week later, a pilot whale stranded and died on Oahu. Five days after that, another pilot swam into the same bay, wobbled into the shallows, and died on the beach. "He looked confused," said a paddler who had seen the whale. "He came right toward us and bumped into our canoe." Two more large black dolphins, identified by onlookers as false killer whales (almost certainly incorrectly, according to Baird, who suspects these were pilot whales too), were also spotted in the vicinity, circling erratically as though in distress; tiger sharks had been sighted skulking nearby.

Were these RIMPAC casualties? Not surprisingly, the Navy denied it. To speculate that sonar or missile tests had been to blame for the dolphin traumas would be "premature and irresponsible," a Navy spokesman said. There was no proof either way, it was true, only ugly coincidence. But it was the same ugly coincidence that had happened so many times before.

≋

The definition of a very good day, for Robin Baird, is one on which somebody encounters false killer whales in Hawaii, and manages to take half-decent pictures of them. That person doesn't even have to be him: it's the sighting itself that is precious. Perhaps the best illustration of how dedicated a scientist must be to seriously study these dolphins is this inconvenient fact: in Baird's eight weeks of boat work around Hawaii in 2014, he and his team crossed paths with pseudorcas on exactly one occasion. But thanks to a flood of photos received from other researchers, whale lovers, scuba divers, ocean aficionados, and sharp-eyed boaters toting Baird's false killer whale field guides, the sum total of the year's sightings was actually 17, with 119 whales captured on film, and 69 individuals identified. When you consider that there are fewer than two hundred pseudorcas left in Hawaiian waters, this was an extraordinary collective effort, a testament to Baird's tireless campaign of public outreach.

But if Baird popped the champagne cork on New Year's Eve to celebrate, he would've been a day early. On January 1, 2015, not only was there a major false killer sighting two miles off the Big Island—a pod of twenty-five to thirty animals, including some calves—but the people who encountered them happened to be a group of professional underwater photographers. The resulting videos and pictures were so lovely, so crystal clear and mesmerizing, that they ended up on the Hawaiian evening news. Watching the footage, I could see why this was Baird's favorite species. The whales whizzed around, below, above, and beside the divers, buzzing them joyfully. It was an astonishing scene: here were two dozen of the world's rarest dolphins, swooping in for their close-ups with intent looks of curiosity and what almost appeared to be jesting smirks on their faces, their sonar clicking and creaking and their whistles ringing through a blue cathedral of sea. The false killers were swift and agile and alert, a breathtaking combination of elegance, power, and obvious intelligence. They were dolphin torpedoes—and, in a perfect world, the only type of underwater missile that we would ever find in the ocean.

CHAPTER 10

CHANGE OF HEART

One blue-sky Saturday morning in May, I rented a car at the Toronto airport and drove south through eighty miles of thick traffic. I passed strip malls, big-box stores, fast-food outlets, office parks, tract housing, and chain motels on my way to one of nature's grandest spectacles: Niagara Falls. I have seen the Falls in their thundering glory, and it is not a vision that ever grows old, but this time I would be stopping a mile short of them to visit another tourist site. Near the turnoff for Rainbow Bridge, the dramatic span that connects to the American border, I saw an SUV with a bumper sticker that said MARINELAND: LAND OF HORRORS, and I followed it. I knew we were going to the same place.

So were hundreds of other people. It was the first day of a new season at Marineland Canada, which meant there would be a protest. On closing day next October, there would be another protest.

It was a cycle that had continued for years. Among many people who cared about dolphins, Marineland had a bad reputation. Lately, the protests had grown louder, and the critics' accusations had been echoed in the media, and the outcry appeared to be headed toward a crescendo. Change seemed imminent—but was it? Canada might sound like an unlikely place to ponder the past, present, and future of dolphins, but two spots on opposite sides of the country offered some startling, and highly unexpected, insights.

When Marineland opened for business in 1961, its owner, a Slovenian man named John Holer, had begun by exhibiting a trio of sea lions in two steel tanks. Now, Marineland's holdings had grown to a thousand acres and the park housed a profusion of animals, including a handful of bottlenoses, about forty beluga whales, and an orca, along with seals, sea lions, walruses, and several forests' worth of bears and deer. Its grounds were dotted by rides with names like Space Avenger, Sky Hawk, and Dragon Mountain. There was also a ride called Orca Screamer, an unintentionally ironic name given the park's history: since the seventies, at least sixteen orcas had presumably died in its care. And the orcas were not alone.

The previous year, the *Toronto Star* newspaper had launched a series of investigative articles about Marineland, prompted by the death of a beluga calf named Skoot. Against all animal husbandry logic, Skoot and her mother, Skyla, had been placed into close quarters with two adult male belugas. At six p.m. one night, while Skyla and a park guide watched in panic—no trainers or veterinarians were on site, and apparently none responded to calls—the males bit and bludgeoned Skoot, killing her. Though the guide chronicled the mauling and his account later became public, Marineland's official stance was that Skoot had "passed away after a sudden onset of illness." Holer acknowledged the attack, but chalked it up to nature's cruel caprices. It was the calf's diseased condition, he claimed, that had provoked the aggression. "You have to understand . . . for people and all living things, there is a time to live and a time to die," he advised the press.

When two *Star* reporters, Linda Diebel and Liam Casey, began to dig, they found a nightmarish trove of stories about Marineland. They were aided by animal activists—who had been pointing their fingers since the seventies—and by numerous whistleblowers, Marineland employees who had decided to step forward in an attempt to help the creatures they'd cared for. The ex-employees told of an occasionally malfunctioning water-quality system that had seared marine mammals with chemicals, among other troubles. Holer repeatedly denied that there had ever been any problems with the water in Marineland's tanks: "We take care of the animals," he told reporters, "better than I would take care of myself."

Some dolphins were housed in a windowless concrete building that was backed by a video arcade; other dolphins had been confiscated by the U.S. Department of Fisheries after Marineland illegally captured them in the Gulf of Mexico. Keiko, the orca in the movie *Free Willy* who had withered in a shabby Mexican marine park, ended up there after Marineland sold him to its owners. In 2011, SeaWorld actually repossessed one of its orcas from Marineland, a male named Ikaika, whom it had swapped for four beluga whales on a breeding loan. Although at least thirty-seven orcas have perished at their own properties, SeaWorld executives were alarmed by Ikaika's living conditions in Niagara Falls. When Holer refused to return the orca, the dispute landed in court, with SeaWorld eventually prevailing.

Marineland's animals didn't fare any better outside the tanks. Bears and deer were spotted with open sores, mangy fur, and ripped ears. In the past, contagious bovine tuberculosis had infected the park's deer; on two occasions, spectators had looked on, horrified, as a gang of bears ripped another bear apart. "Once an animal turns on another, there's not much you can do," Holer lamented after the second fatal bear brawl. "Nature can be awful cruel." One whistleblower, a former maintenance supervisor, revealed the existence of four mass graves on the property, unpermitted pits that were estimated to hold more than a thousand carcasses.

Before I saw Marineland, I heard the protesters. They lined the road in front of the property, waving signs. HONK IF YOU LOVE DOLPHINS, read one banner, and the passing cars leaned on their horns in solidarity. The crowd—later estimated to be a thousand strong—was vocal and impassioned. I found a parking spot and pulled in, next to a woman and her nine-year-old son. The boy carried a sign that said WELCOME TO SLAVELAND. "I feel bad about the deer," he told me.

Almost immediately I spotted Ric O'Barry, wearing his trademark baseball cap and Ray-Bans, surrounded by people. We greeted one another, and he spelled out the problem: Canada had approximately zero regulations to protect captive marine mammals. "You could literally dig a hole in your backyard, fill it with your garden hose, and put a dolphin in there," O'Barry said. "It's completely legal in Ontario."

Absurdly, this was true. O'Barry's assessment was, if anything, understated. Here, you could have a beluga whale and a tiger in your backyard too, should you happen to want them. Countries like Bolivia, Chile, Costa Rica, Croatia, Cyprus, Hungary, India, Slovenia, and Switzerland had ended dolphin captivity on moral grounds; others, including the U.K., Luxembourg, Nicaragua, and Norway had curtailed the practice to the point of oblivion, and the U.S. had its venerable Marine Mammal Protection Act. But Canada offered nothing, despite its progressive record on so many other issues. Marineland's regulators were so toothless as to be pathetic.

With the government idle, the public had taken up the cause with a grassroots fervor. As we stood next to a wire fence separating protesters from the patrons who were buying tickets to Marineland that day, O'Barry introduced me to Cara Sands, a documentary filmmaker from Toronto who had been trumpeting the park's inhumane practices since 1990. Sands, a raven-haired, doe-eyed woman in her forties, had crawled through a boiler room and shot clandestine video of Marineland's holding facility, an off-limits bunker known to trainers as "the Barn." "I found this warehouse full of dolphins," Sands recalled, "and one killer whale named Junior. I tracked the situation for five years." She pointed over the fence to a long, slablike building. "It's right there. That's where Junior was held."

Like Keiko, Junior was an Icelandic whale, taken from his pod in 1984. His exact age at capture was unknown, but he was so young that the first time Sands saw him he still had vestigial hairs on his face, a feature orcas lose early in life. On two other occasions Sands returned to the Barn; both times Junior floated in residence, visibly fading. "Was he ever brought outside?" Sands said. "Not that I'm aware. Not that I've ever seen. I can't prove it, but I think it is possible he was in that building for four years." Junior died in 1994. Marineland has never commented on Junior's life in the warehouse or confirmed his death.

Having lasted a brief, miserable decade, Junior was an elder compared to many of Marineland's deceased orcas. Kandu II, a male snatched from Washington State, died at age eight. Hudson, a captive-born orca, reached six; Nova, Kanuck, Malik, and Athena were all around four. Algonquin survived for two years and eight months. Two unnamed calves died at three months old. April was gone in a month. Another anonymous calf lived for eleven days.

Over the years, twenty-nine orcas had ended up at Marineland, and now only one remained. This was Kiska, who swam alone in her pool. During her years of captive breeding she had delivered four calves here, and lost them all. In 2008 she also lost Nootka V, an Icelandic matriarch who had been her closest companion. Nootka V herself had lost eight calves, and she and Kiska had supported one another through their multiple births.

Wandering through the gathering, I met Christine Santos, Kiska's former head trainer. Santos, a slender thirty-four-year-old with a pensive air, had been fired after her boyfriend, Phil Demers, another senior trainer, resigned to become a whistleblower. When Santos was let go, she expressed concern to the *Star* reporters that Kiska's tail had been bleeding profusely. Diebel and Casey found a video showing exactly this. Now, both Santos and Demers were being sued by Holer—who made a practice of serving SLAPP suits on anyone who spoke out forcefully against Marineland—and in turn they were countersuing, claiming Holer's lawsuits were an abuse of process. "I have to worry about all this legal stuff," Santos said. "But I can tell you how I feel about Kiska."

Santos had looked after the orca for the past twelve years. "It's kind of surreal for me to be on this side of the fence, and not with her," she said, her voice shaky. Santos described Kiska as the "sweetest girl ever," an inquisitive orca who liked to play hide and seek, and have her tongue rubbed. In the past, Santos had lugged a television down to the underwater viewing window; during winter she built snowmen around Kiska's tank, while the whale spy-hopped to watch. "Anything to keep her stimulated," Santos said. *"Anything."* She began to cry.

≈

Orcas are not casual animals. At a glance they all look alike, but there is nothing homogenous about them. In many cases, orca clans are so distinct from one another that scientists suspect they are actually separate species. In the Pacific Northwest alone, there are three types of orcas, commonly referred to as residents, transients, and offshores. Killer whales from each of these groups have about as much in common as investment bankers, rock stars, and nomadic herdsmen. The residents prefer Chinook salmon, choosing to go hungry rather than settle for, say, sockeye. The transients hunt seals and sea lions and won't touch salmon of any kind. Offshore orcas like to dine on sharks, expertly filleting the liver. There is no overlap between these tribes, and they actively avoid one another. If they do make contact, there can be scuffles. When two pods from the same clan meet up, however, they may greet each other in an elaborate ceremony.

To our ears, orca vocalizations sound like a keening ghost choir, unearthly cries that all meld together, but scientists analyzing their calls have discovered that each resident pod has its own dialect. (Likely this is important for mate selection: in the wild, orcas are careful not to inbreed.) They also have unique styles of communication. Residents vocalize loudly while they hunt; transients are stealthily silent, busting out gleefully only after they've fed. Offshore orcas slap their tails while they are swimming, for reasons nobody knows. All three groups wield their echolocation with individual flair. Even their striking markings are not a one-size-fits-all orca uniform: across the oceans, the details of killer whales' body sizes, fin shapes, and white patches vary widely.

As recently as the sixties, orcas were viewed as bloodthirsty marauders—thus, their intimidating name: killer whales. The larger males were dominant, people believed, and kept their females in a harem. Surely these monsters would attack if given the chance; clearly they posed a lethal threat to any person—or even boat—that crossed their paths. They were, in the popular mind, "the biggest confirmed man-eaters in the ocean." Those who encountered orcas often expressed their fear with bullets: whalers, fishermen, naval officers, and air force pilots used them for target practice. At the time, of course, absolutely nothing was known about these animals, aside from the observable fact that they were good at catching their food. But soon, as marine scientists began to study them, they began to realize how wrong the orcas' menacing image was.

Despite ample opportunity and provocation, orcas have never killed a single person anywhere in the world's seas. (By comparison, in 2014 dogs killed forty-two people in America alone.) They are apex predators who show gentle curiosity toward us, expert communicators who demonstrate complex wisdom. Far from being stuck in a harem, female orcas control their pods—especially the oldest grannies. These four-ton ladies are the suns around which all other killer whales orbit: throughout most of their lives, orcas stick close to their mommas. An orca's pod is his immediate family, a group that might contain four generations. Each pod, in turn, is part of a bigger clan, also comprised of close relations, and at the top of the orca organizational chart are the broader communities, made up of clans. In our terrestrial world, we call them nations.

A killer whale's cultural education unfurls at a pace that is similar to ours. They develop socially, like we do, and nothing happens instantly: maturity comes over time. At twenty, orcas are still learning and growing. Matriarchs have been known to approach the century mark, often outliving their offspring; they can spend more than thirty years in menopause. As with all of nature's successful adaptations there is a reason for this, and we've learned some eye-opening things about the matriarchs' role. They babysit, for one thing. They share food. And they teach: killer whale mothers, grandmothers, and great-grandmothers pass on so much essential knowledge that calves removed from their influence are as ill-equipped for wild orca life as children raised by wolves would be in our society, if dropped into Midtown Manhattan.

This blue orb we live on? It's had oceans for 3.8 billion years. In a realm where history isn't written, it is the matriarch who carries the past, everything her pod needs to survive into the future. She is the keeper of the dialect; she teaches her descendants their very identity. She shows everyone how to hunt, no minor task when you consider the tricky and specialized techniques orcas deploy. In Argentina, one group uses the high-stakes tactic of intentionally stranding themselves—rocketing onto the beach, grabbing a seal, and then hopefully propelling themselves back into the water. Scientists have watched young orcas being tutored for six years before they even attempt this.

Depending on where he lives, an orca might learn to blow intricate bubble curtains that herd herring, or pop his head above water to identify a delectable species of seal. He may need to understand fast-moving formations that would humble the Hogwarts Quidditch team, working with his pod in a three-dimensional life-or-death match against sperm whales. He might practice the deft extraction of sting-rays from a muddy seafloor (without getting stung), or be shown how to immobilize a great white shark, restraining it until it drowns. None of these are moves for the uninitiated.

The matriarch also knows *where* to hunt, and when. As the climate swoons and soars, as our tinkering becomes increasingly apparent beneath the waves, the ocean can tilt radically. In an El Niño or a La Niña, for instance, or when cycles like the Pacific Decadal Oscillation leave one year's flush hunting grounds barren the next, or when an orca clan that survived for centuries on Patagonian toothfish suddenly finds longlines raking the waters instead, the matriarch is the key navigator. Using everything she knows, she'll figure out Plan B.

When describing the many sublime characteristics of orcas, even the most empirical scientists can become emotional. Marine ecologist Robert Pitman summarized the feelings of many of his colleagues when he appointed orcas "the most amazing animals that currently live on our planet." Another scientist appointed them "the unchallenged sovereigns of the world's oceans." Academic writing isn't known for its exuberant use of exclamation points, but reading orca research papers, I came across more than a few. Among marine cognoscenti this kind of enthusiasm is expressed often for killer whales, in language usually reserved for five-star movie reviews.

~~~

After mingling among the protesters for a few hours, I decided to venture to the other side of the fence. Marineland's parking lot seemed to go on forever, a desultory hike across a vast expanse of blacktop. I walked in behind a family of five, parents with two young boys and a toddler. They glared at the crowd outside the gates. "Get a job!" the father yelled at the protesters. "Take your Ritalin!" one of the boys chimed in. The other boy busied himself by sighting his fingers like a rifle and aiming at songbirds flying overhead.

The cementscape continued inside the park, done up in an appropriately medieval theme. Marineland sprawled for acres: to find out firsthand how forty beluga whales could possibly be jammed into one man's tanks, I would have to wind my way through many other exhibits.

"Deer Park" and "Bear Country" were equally grim, the animals massed in dirt courtyards or on scrub lawns. A tired-looking senior citizen with some missing teeth stood in a nearby kiosk, selling sugar cones full of honey-coated corn puffs for $2.75 to people who wanted to feed the bears. "Give them the cone, too," she advised patrons. "They like the honey." As the cereal rained down on them, a dozen bears sat around a concrete moat, looking defeated.

Then, the belugas. The many, many belugas. I have never seen these animals in the wild, but I think it is a safe assumption that out there they spend at least some of their time swimming underwater. Here, they hung like buoys, in a vertical position—the better to fit them all in, I supposed—hugging the perimeter of their tank, their heads out of the water and mouths agape, begging for little fish ($8.75 a bucket) tossed to them by children. Every so often a trainer would clamp a beluga's mouth shut for a moment so a kid could fondle his head. This may have been a newly adopted precaution: in the past, the beluga-feeding public had been bitten. And as recently as 2006—after Tilikum had well established the worst-case scenario—Marineland had encouraged visitors to touch and hand-feed its orcas.

I stood back and took in the scene. Belugas are lumpen in the most adorable way. When you see them, you just want to cuddle them. Their bodies are long, tapered at both ends but concave in the middle. Their heads are bubble-shaped with short, triangular beaks and jellybean eyes. They lack dorsal fins, which is ideal for slipping under ice: most belugas live in arctic waters. They are especially expressive whales, known for their chattering chorus of chirps and creaks and buzzes and songs. Belugas are also accomplished mimics. They have been known to playfully imitate the noise of passing motorboats; one captive group that could hear a nearby subway began to make subway sounds. Their faces wrinkle and emote in ways that seem delightfully familiar to us—they remind us of more innocent versions of ourselves. (Their closest relatives, however, are the exotic narwhals, toothed whales whose single, skewer-like tusks bring to mind unicorns.) In the dark northern waters where belugas romp, their spectral white bodies stand out like a visitation. Only the hardest heart would fail to be charmed by them.

Marine park owners know this. The belugas' charisma has made them the hot new draw in captivity. But the demand for belugas far exceeds the supply. For years they were hunted relentlessly, and they are still being plucked from the waters around Russia, but the planet's remaining belugas grapple with an even more widespread threat—pollution. They are somewhat fragile animals, prone to illness and cancers, extra susceptible to the polycyclic aromatic hydrocarbons, organochlorines, PCBs, dioxins, lead, and mercury swirling in their waters. Scientists studying belugas in the mightily contaminated St. Lawrence River have seen evidence of widespread miscarriages, still-births, and calf mortality: In 2013, a record seventeen dead calves were found drifting in the currents. One recent study of twenty-four St. Lawrence belugas turned up twenty-one tumors.

Belugas aren't robust captives, either. In the wild, a healthy animal might live to sixty; in the tanks, it's the rare beluga who makes it past twenty. It is impossible to establish a firm number of how many belugas Marineland has kept or lost—there have been so many of them, the record is fuzzy, Marineland hasn't commented and Canada has no official registry—but one source, Ceta-Base, which monitors marine parks worldwide with a heroic level of detail, tallies the park's total population (since 1999) at seventy. This is some ugly math. According to Ceta-Base, Holer has taken delivery of thirty-five wild-caught Russian belugas. On top of that, the site also lists thirty captive births and four stillbirths, which, if accurate, meant that some twenty-nine belugas have ended their lives in Niagara Falls. But there was a note at the bottom of Ceta-Base's roster, written in bright red letters: *This list may be incomplete.*

≋

"Why doesn't he splash more?"

The child was whining, bored by Kiska, who swam by so slowly it was as though she were conserving energy. "This is boring," the boy said, tugging at his mother's fanny pack. "Let's go."

I suppose the good news about Kiska was that on this day she didn't appear to be bleeding. Her dorsal fin drooped like a deflated balloon and she showed no interest in her surroundings—not the people, or the beluga-palooza going on next door to her, or the three rocks scattered on the bottom of her tank. There was nothing else. The orca circuited on autopilot, making no sounds, her dirigible of a body suffused with a quiet grace.

To me, Kiska looked like depression in motion, but Marineland claimed that she was the luckiest of creatures. "Kiska is now quite elderly," the park's public relations stated, "and like all elderly animals or people, choses [sic] to do things at her own pace, in her own time, and as she pleases." No one should think this whale was anything but pampered, Marineland asserted. "Kiska receives continuous care and treatment, if necessary [sic] that your average person could only dream of." At thirty-eight, Kiska was only middle-aged, but the life she led now would make anyone old.

On a cruelty scale, keeping an orca in isolation ranks near the top, right up with the solitary confinement of a person. Prisoners quickly implode when locked up alone: it's a fast path to complete mental meltdown. Humans are fundamentally, biologically, evolutionarily social creatures—and so are killer whales. If anything, they're *more* social than we are. The world's only other lone captive orca, a female from Washington State named Lolita, lived at the Miami Seaquarium, where, for the past forty-three years, she had circled her tiny pool like a goldfish in a shot glass. To view aerial footage of 18-foot Lolita looping endlessly in her 35-foot-wide tank is to despair on a cellular level. Like Marineland, Seaquarium was the site of loud and frequent protests.

I went below to watch Kiska through the underwater window. The viewing area was low-lit like a grotto, the tank casting a cool blue gloom. The space was packed with people, stuffed with strollers. Kiska glided by regally, but there was no joy in her movements. She was simply adrift in a lonely ether. I looked on for a few more moments, snapped some photos, and left, feeling low. Later that afternoon I had a plane to catch; I was flying to Vancouver, spending the night, then continuing on to Victoria. From there I would catch a ferry to Pender Island, in the home range of Lolita's clan, the Southern Residents. These days, it was also the home of Chris Porter.

≈

Pender Island is the land of rain and giant trees, of water and wood and mildew. Slipped between Vancouver Island and British Columbia's Pacific coast, this fourteen-square-mile island is a mere punctuation mark. But for Chris Porter, Pender was something far bigger: it was a staging ground, the place where he would plot his redemption.

Having seen the results of interviews Porter had given during his Solomon Islands years, I had wondered if he would even agree to talk to me. How many times, after all, can a person be referred to as "the Darth Vader of Dolphins" and still want to continue the conversation? But when I contacted him, Porter was not only willing but eager to meet. In fact, the openness of his response startled me. Even people with routine secrets tend to be cagey, but here was Porter, freighted with controversy, with years of dolphin trafficking he seemed happy to discuss.

But there was a likely motivation behind Porter's availability. In e-mails we'd exchanged and articles I'd read, he'd described some heavy soul-searching that had led to a surprising decision: he had renounced the captivity trade. "I'm disillusioned with the industry," he'd said. "I'm just tired of it. I have realized there are other ways to educate people about the importance and intelligence of whales and dolphins without separating them from their family groups." This was a 180-degree twist, of course, coming from a man who had sold eighty-three dolphins into captivity in recent years. Porter acknowledged his past—"To be sure, I have a bad name"—but pledged to now shift his attentions from selling dolphins to saving them.

When giving reasons for his about-face, Porter cited a number of sorrowful stories. He had been moved by certain dolphins in the Solomons, he said, and he'd felt that what he was doing to them was wrong. But the animal who had affected him most, the one he returned to again and again, was Tilikum. It was Dawn Brancheau's death that had jolted him, Porter claimed. This was the final straw for a personal reason: twenty years ago, at Victoria's Sealand of the Pacific, during Tilikum's early years as a captive, Porter had been one of his trainers. When SeaWorld bought the orca in 1999, in the wake of Keltie Byrne's death, Porter was among those who had helped transfer him to Orlando. He knew Tilikum, he knew him well, and now this whale had proven emphatically that the system was broken. To hear Porter tell it, whatever he did next, he was doing it for Tilikum.

Was his conversion genuine? Perhaps. At the moment, Porter was still in exile, still mopping up his past. These days he worked as a night auditor at Poets Cove, a resort and marina on South Pender's peaceful, picturesque coastline, and that is where we would meet.

≈

"I never knew it would get as big and complicated as it got. Yeah . . . I was surprised at how complicated it got. At the end, I was just . . . feeling like I was drowning."

Porter took a sip of his lager. He looked drained, and somewhat morose, but his blue eyes were sharp and wary. His brow, fully revealed by a receding hairline, furrowed when he spoke. We were three hours into an interview that couldn't have been much fun for him, but I kept asking questions and he kept answering them. I wanted to know what had propelled him to the Solomons in the first place, why he made the choices he did, and if he regretted them. I wanted to understand how a person who claimed to love marine life could become a dolphin trafficker in the first place.

"I did it to challenge the industry," Porter told me. "Like, let's be honest about where we're getting these animals. If we can't even do that, then how can we say we're educating the public and making them aware? If we're doing something really good, then why do we have to hide where this dolphin is coming from? Why can't we celebrate where it comes from?"

I listened, though it struck me as unlikely that the more details the public learned about the highly profitable dolphin trade, the more they would embrace it. In my experience, marine park–goers were willfully oblivious to the animals' provenance, and the marine parks encouraged this. The entire industry, in fact, was built on the illusion that the dolphins had somehow dropped from the sky. But I suppose this was Porter's point: if you can't be transparent about what you're doing, then perhaps you shouldn't be doing it. His goal in the Solomons, he said, had been to create a paradigm where everyone—dolphin-hunting tribesmen and dolphin-loving tourists and the dolphins themselves, and of course, Porter and his investors—would have benefited. Though, as he knew, it hadn't exactly worked out that way.

"I thought, 'Why not make a resort?'" he explained, in a torrent of words. "And you could have a big bay and you could have guests coming down at midnight because dolphins are more active at night—it's artificial that we make them do shows during the day, because that's when the people are there—and so you get shows at night, even just watching them at night's better because they're active, they're playing, they're mating, you know. It's a fun time." He paused for breath. "So I thought, wow. I've done a lot of open-ocean work in my life. I know how to . . . You could have the dolphins going and coming back—at breakfast, you could open up the blinds and there's a dolphin meeting you."

In Porter's scenario, the dolphins would stay only as long as they pleased. They would be fed regularly, but not enough to let their hunting skills grow rusty. Ideally, the dolphins would come and get their fish snack, then play with some guests in the lagoon, becoming regulars even in the absence of nets. "Some of the animals I've worked with open-ocean, they leave," Porter elaborated. "They're just like—'Bye! I'm outta here.' Others are happy to stay for a bit."

This plan began to percolate in 2000, when Porter left his job as head trainer at the Vancouver Aquarium and moved to Genoa, Italy, to work with the dolphins at that city's facility. His wife and three kids had adored Italy, but Porter was itching to go bigger. Though he'd never heard of the Solomon Islands, when someone mentioned its dolphin-hunting tradition to him, Porter was intrigued. "I figured, 'Hey, if they're willing to kill them, then they're probably willing to let me keep some alive. So I went down there and started meeting the locals and found a cool spot to set it up: Gavutu. But as soon as I came back all the investors said, 'No way! That place is crazy!'" He chuckled at the memory. "That's when I approached the government to export a few—so I could self-fund the resort."

I reached for my beer. Porter's interpretation of events was highly debatable, including the fact that eighty-three dolphins were more than "a few." How did he feel, I asked, when his plans for Gavutu failed? "It's taken me a while to get to grips with it," Porter admitted. "I went through a rough time. For the past two years I was off the rails, and I was probably off the rails in the Solomons, too, trying to deal. You know, it was an intense situation." He paused, and stared into the middle distance. A long moment ticked by, as he gathered himself. Then he continued: "I'm at the point now where I say okay, if one individual caused that much damage, then one individual can cause that much good."

"What damage do you regret most?"

Porter stopped short, as though startled by the directness of the question. He paused again, but this time the tears overwhelmed him. After a moment, he exhaled. "To individual animals," he said, almost sobbing. "Yeah, it's the individuals. Free the Pod died, they all fucking died, right down to the last baby, and you know . . . it hits me now." He cried some more. "Sorry," he said, wiping his eyes. "*Whew.* I knew this was gonna be . . ." His words trailed off.

Ultimately, the sincerity of Porter's remorse will be evident by his actions, not his tears—but when he outlined his new project, everything about it sounded positive. Using new technologies, he wanted to immerse viewers in ocean simulations; to let people virtually ride along with a pod of orcas, for instance. As he spoke, Porter became animated, rattling off his plans with entrepreneurial zeal. Maybe the simulator would become a ride. Or maybe it would consist of a module that people could drive around below the sea surface. "Like a diver bike," he said. "Have you ever seen those diver bikes? They're from Germany. They can do eight knots underwater!"

These ideas were already in play, Porter told me, elsewhere around the globe. "I've heard Richard Branson is talking about doing simulators now. Good! And then there's James Cameron. Good! I really believe we're in a moment when innovators getting together on behalf of the ocean—we'll be able to make marine parks archaic." If this vision were realized, then one day soon we would look up and no kid would even *want* to go to a Marineland or a SeaWorld or a Shenzhen Ocean Safari Park or a Janohire Dolphin Farm or a Yaroslavsky Dolphinarium, unless these places had evolved into something completely different: businesses that truly benefited the natural world.

Porter had indicated that he would be shooting the simulator footage himself, swimming first with sea lions, to experiment. "How will you film the orcas?" I asked. "Will they let you get that close to them?"

"Well, I'll be honest," Porter replied, with a sly smile. "Because I know how to catch them, I know how to get them on camera."

≈

During the sixties and early seventies, a brutal era of orca captures, the Southern Residents who swam past Pender Island had been targeted more than any other group. Orca hunters, working for marine parks, had used spotter planes, explosives, and nets to herd the pods, injuring and killing many animals in the process, traumatizing them all, and in the end removing forty-five individuals, 30 percent of the Southern Residents' population. Worse, it was calves the marine parks wanted, so they'd effectively destroyed the next generation.

Responding to public outrage, Washington State sued SeaWorld in 1976 for overstepping its collecting permit. The state won, and then proceeded to outlaw further orca captures. (SeaWorld immediately moved on to Iceland and began to capture whales there, a practice it continued until 1989.) Since then, the Southern Residents have struggled to recover, but there are only seventy-nine of them left in an area rife with overfishing, sonar use, missile testing, shipping traffic, and pollution. The group is now listed as endangered.

The sole surviving orca from those shameful roundups was Lolita. After forty-three years in a tank she stood no chance of reintegrating fully into the wild, but a movement had gathered steam to relocate her to a natural sea pen: Kanaka Bay, on Washington State's San Juan Island, near the place where she was caught. Compared to the Miami Seaquarium, Kanaka Bay would be a vast improvement, a taste, at least, of freedom. Even better: Lolita's mother, an orca from the Southern Residents L-Pod, was still alive. The matriarch was eighty-five now, and traveled with her family. Scientists from the nearby Center for Whale Research spotted her regularly. It was possible, even likely, that Lolita would hear her calls and recognize them.

After the dark ages of hunting and trafficking dolphins, after all the pain we've caused them, there is a glimmer of hope—fast becoming a gleam—that the future might be different. Porter's change of heart is only one example. The documentary *Blackfish* has followed in *The Cove*'s wake, opening millions of eyes to the real cost of keeping killer whales, animals who could never be content in even the most immense concrete tanks imaginable. As a result of increased awareness, California and Washington State have both proposed new laws banning orca captivity and breeding, with other states likely to follow. In 2014, the National Aquarium in Baltimore, Maryland, announced its intention to create a seaside sanctuary for older captive dolphins, a kind of dolphin retirement home. Such places already exist for elephants, chimpanzees, and other animals, CEO John Racanelli pointed out to the press. "We want to do right by our dolphins and by our audience, and do a better job of serving our mission," he explained. "We want to change the way humanity views and cares for the ocean." The National Aquarium has also terminated its scripted dolphin show, deeming the practice "antiquated."

For marine parks, adopting a more enlightened outlook on the subject of captive whales and dolphins is clearly the way forward. Seaworld's high-flying stock cratered shortly after its IPO, signaling the public's growing distaste for its operations. Since May 2013, the company has lost more than a third of its value, and is now being sued by its shareholders for "failing to disclose improper practices regarding its orca population and denying the cause of the company's failing attendance."

On another front, an organization called the Nonhuman Rights Project, led by lawyer Steven Wise (with Lori Marino on board as its scientific advisor), was beginning to press cases asking the courts to give certain brainy animals—like chimpanzees, elephants, and dolphins—some basic rights under *habeas corpus,* which grants status to captive persons. This process is arduous, pioneering, and sure to face resistance, but Wise's approach is clever. Legally and philosophically, the term "person" merely indicates a distinct entity; it is not synonymous with "human being." After all, if corporations are considered persons under the law, then why not creatures who demonstrate such sentience and self-awareness that they call themselves by name?

This same idea had bubbled up at the 2012 meeting of the American Association for the Advancement of Science, the world's largest convocation of scientists, where one group presented a ten-point "Declaration of Rights for Cetaceans." "No cetacean should be held in captivity or servitude; be subject to cruel treatment; or be removed from their natural environment," the declaration read. "No cetacean is the property of any State, corporation, human group or individual." "Dolphins are nonhuman persons," ethicist and philosopher Tom White explained. "A person needs to be an individual. And if individuals count, then the deliberate killing of individuals of this sort is ethically the equivalent of deliberately killing a human being."

Now the gleam was visible to the naked eye, and still the good news kept coming. After my meeting with Porter, I didn't hear from him for a while. When I did, it was for a good reason: his project, Ocean Walls, had gone live in a British Columbia mall. Porter sent me a video showing the exhibit in action, with dozens of people standing in front of its high-definition screens, mesmerized by the Southern Residents. "For Tilikum," the title read, as the images appeared. A BEAUTIFUL ALTERNATIVE TO MARINE CAPTIVITY, praised the headline on TheDodo.com, an animal lovers' Web site. Porter had also organized marine conservation talks, held against a backdrop of orca footage.

Beluga whales got some welcome news when the National Oceanic and Atmospheric Administration denied a petition from the Georgia Aquarium to import eighteen wild-caught animals from Russia. After lengthy hearings, NOAA ruled that removing these belugas would contribute to the population's decline; that these captures would likely lead to more captures; that the aquarium had not adequately proven the whales had been treated humanely, and that, all in all, it was a bad idea.

And then, a surprise: eighteen months after I'd visited Marineland Canada, the Ontario government issued a 125-page report, written by a team of scientists, laying out sweeping protections for captive marine mammals. Also, it advanced legislation prohibiting the future acquisition, sales, or breeding of orcas. Since Marineland was the *only* facility in the province that kept any of these animals, one could say these lovely new rules were written especially for John Holer.

"I envision the day when the current oceanaria will progress from being 'prisons' for dolphins to being interspecies schools," John Lilly wrote, "educating both dolphins and humans about one another." Like so many of Lilly's predictions and pronouncements and hopes, the idea of humans and dolphins thriving together is worth dreaming about—but maybe now we were actually getting there. One might imagine such a relationship to be unprecedented, except that it isn't. It has happened before, in a time so long ago that it almost seems like a fairy tale.

CHAPTER 11

THERA

The dolphins were 3,800 years old but still lively, their black and white bodies arcing through terra-cotta waves as they leaped across the prehistoric pottery. They regarded me with their oversize feline eyes, outlined with kohl in the manner of Cleopatra. But these dolphins preceded that Egyptian queen by seventeen centuries. They were among the earliest dolphin images known to us, the first representations of their kind, painted on what looked like a rough clay breadbasket. The artist was nameless, the civilization he belonged to shrouded in mystery, but one thing was obvious: whoever created these dolphins had done so adoringly.

I stepped back so I could see them in the spotlight. If you didn't know the origin of this piece, you might find it underwhelming. It was, after all, only a kitchen utensil. The National Archaeological Museum of Athens was the territory of thunderbolt-hurling, larger-than-life Greek gods, not whimsical little dolphins. Learning the dolphins' story, however, would change your perspective. This pod's place in history was central, and the tale that accompanied them one of humanity's greatest riddles.

I'd arrived in Athens the previous day, landing in a doleful city wrung out by debt and austerity. No one was thinking about marine life. They were knotted with worry, wracked by unemployment and uncertainty about the future. My cabdriver, a Greek man in his thirties, told me that he had been trained as a civil engineer. He bemoaned the state of affairs that had landed him behind the wheel. "I think by this time next year we will be eating one another," he said.

The streets were careworn and grimy and heavily graffitied. Every other storefront was empty; stray dogs prowled the streets. Even in the museum, one of the world's finest collections, here and there the lights were burnt out, giving the building an air of abandonment and neglect. I walked through arcades of magnificent torsos, past marble friezes depicting glorious battles, memorials to classical Greek history—but I had come to see the dolphins. They were tucked into a small, underlit room on the second floor, the Thera Gallery. You could easily miss it, which was somehow fitting: the Minoans, the people who had painted these creatures, eluded us for over three thousand years.

At the height of their civilization, around 1700 BC, the Minoans had lived among the Cyclades, a group of islands in the Aegean Sea. Their main strongholds were Crete and Thera (also known as Santorini), but their influence was felt all the way to Asia, Africa, and possibly beyond. They were master seafarers, ocean savants, celestial navigators who had no qualms about sailing over the horizon at a time long before maps. Their nautical adventures were motivated by trade: luxurious Minoan goods—sumptuous wall paintings, fine pottery, gold jewelry, bronze tools, olive oil, an intoxicating drink they'd invented called wine—were coveted by even the Egyptian pharaohs. As a result, the Minoans were wealthy, and they built palatial complexes that were dripping in beauty. They were confident, too: unusually, none of their settlements were fortified. As a cosmopolitan people with roots that can be traced back to the Stone Age, you'd think the Minoans would be as familiar to us as the Egyptians, the Romans, the Persians, or the ancient Greeks. Instead they had vanished, their existence one long string of question marks.

Perhaps it's easy to disappear when you're buried under a hundred feet of ash, which is how their culture met its demise. Sometime around 1500 BC, an epic volcanic explosion on Thera snuffed out the bright spark of Minoan life. This cataclysm was unlike anything in history, estimated to have been four times more violent than Krakatoa, the Indonesian volcano that erupted in 1883, partially collapsed, and killed 36,000 people. When Thera's volcano blew its top, it belched clouds of smoke and gas and rock miles into the air. It blotted out the sun, darkening the sky for months. Its roar was heard in Africa; its ashes fell in Asia. The island's peaks heaved and shuddered, cratering into the sea, kicking up a towering tsunami that would have galloped over to Crete in less than half an hour. Where there had been mountaintops, now there was a caldera of ocean 1,200 feet deep. Once-grand palaces and cities sank beneath the waves—many researchers believe that Thera inspired Plato's mythical Atlantis. No one knows exactly what happened to the Minoans after that, but by 1400 BC they were all but gone from the scene, replaced by the warlike Mycenaeans.

The Minoans had been erased so thoroughly that it was only by chance that archaeologists unearthed them, and to date we've glimpsed only a fraction of their heritage. But what little we do know about them, gleaned entirely from the artwork they left behind, suggests a happy, sophisticated, nature-worshipping people. Among thousands of Minoan paintings and ceramics and carvings, many of them meticulously detailed, there is a distinct absence of imagery portraying war or fighting or any kind of violence. Their god was, in fact, a goddess: the Minoans revered a figure called Potnia, an icon of the divine feminine. She was the source of love and creation, and the mistress of the animals: in exquisite frescoes and gold seals, she is shown dancing in olive groves with her attendant priestesses, with lions and griffins at her heel, with birds wheeling overhead and dolphins cradled in her arms.

Among the most sublime Minoan artifacts was a series of frescoes found at Thera, in a lost seaside city known as Akrotiri. Greek archaeologist Spyridon Marinatos discovered the site, digging on a hunch in 1967 and instantly hitting pay dirt. Akrotiri turned out to be another Pompeii, only 1,600 years older. Peeling away layers of earth, Marinatos uncovered two- and three-story houses with amenities nobody expected to find in a prehistoric settlement: the Minoans even had running hot water. In the prelude to the volcanic apocalypse, the people had fled—absolutely no human remains have turned up—but the buildings' walls were covered with murals, done in a distinctive, evocative style. They featured lavishly adorned women bending to caress lilies, swallows kissing, dolphins and monkeys frolicking, spirals painted in kaleidoscopic colors, and ceremonial gatherings attended by blissed-out Minoans, all drawn with a lyrical grace. Marinatos also found many Theran possessions that had been left behind, and the dolphin pottery at the museum in Athens had come from his excavations.

Taken as a soulful snapshot of a time and a people—what they cherished, how they lived—the Minoan artworks tell a spellbinding story, with dolphins among the key players. Other cultures have been drawn to dolphins—few animals have been mythologized so thoroughly—but the tales have been passed down orally, so they seem hazy and unreal. Their narratives, while enchanting, are hard to swallow as nonfiction. The West African Dogon people, for instance, claim their forefathers were dolphin-like beings called Nommos, who descended to earth from Sirius, a star system in the constellation Canis Major. In the Amazon, there is widespread belief that the river's pink dolphins, or botos, are clever sorcerers who often appear on land disguised as handsome men, bent on seducing women. The botos are envoys from a parallel universe, the shamans say, guides between this world and an underwater realm called the Encante, a crystal metropolis where everything shines like diamonds. The notion that dolphins can morph into humans whenever it suits them is also found in Aboriginal Australian legend, Pacific Island chants, Native American folklore, and Greek epic poems. Certainly these myths, true or not, reflect an incomparable relationship between our two species. But while other cultures talked about their bond with dolphins, only the Minoans provided proof.

Dolphins show up with startling frequency in their art, so often that historians refer to this as the Minoans' "marine style." If there were any people who painted dolphins earlier, or more often, or more brilliantly, or exalted the animals more, we haven't yet found them. So who were these playful, peaceable dolphin lovers—and what are they whispering to us from millennia ago? These aren't idle questions. In an age when we coexist uneasily with every other life form, it seemed to me that we could use a little Minoan wisdom to go along with our bottomless stores of information.

Outside the Thera Gallery, in the rest of the museum's displays, I was thrust into a fiercer, more recent era of warrior heroes and terrible gorgons, satyrs and pygmies and lions tearing apart horses. Zeus was spearing women everywhere I turned. As the centuries progressed, the dolphin images dwindled, unless the animals were sidekicks to a gruff Poseidon, rearing up with his trident. Everyone had weapons, many bodies were missing their heads, the ocean was seen as a hostile combatant, and no one looked like they were having nearly as much fun as the Minoans.

≈

The old man, Artemis, sat at a rickety table and sipped from a glass of honeyed raki, his back leaned against the wall of his whitewashed house. The sun was setting over the caldera and people had gathered along pathways, in doorways, on promontories, to watch its blazing descent. All of Thera faced the sea, its buildings clinging to sheer cliff faces that plummeted to the water. In the gilded dusk, the island's snow-white architecture was bathed in molten gold and copper. Artemis was pushing ninety and he didn't speak a word of English, but that didn't impede our conversation. He gestured to the glowing landscape and touched his heart. I nodded, and did the same. Then I drew a stick-figure dolphin on a scrap of paper and handed it to him. "Ah," he said, "Akrotiri!" He picked up the raki bottle and refilled my glass.

There was nowhere I'd ever seen that had half the drama of this blasted-out island, nothing that stirred my imagination more. Motoring in on the ferry that morning, I'd reeled back in my seat when Thera came into view. Where there had once been an almost perfectly round landmass, now there was a ravaged crescent of volcanic rock banded in striations of rust, brown, and gray. It was as though a chunk of the place had been clawed away by a giant paw and flung into the abyss. Inside the caldera, the water lay as still and inscrutable as black marble.

I had immediately set out to explore, walking narrow paths shared by sure-footed donkeys. If you slipped here, you would fall for a mile. In the pretty town of Oia I found dolphins everywhere: stenciled on walls, stamped onto hotel signs, etched into jewelry, inked on pots that were replicas of Minoan designs. In one store, I picked up an amphora ringed by cavorting dolphins, and caught the owner's attention. He reminded me a lot of the Cat in the Hat. "Yessss," he said, in a smooth Greek accent, "the symbol of the Minoans. They were very preoccupied with marine life, you know."

Continuing on, I had come across Artemis, who looked like he might have been here long enough to have personally known some Minoans. *"Kalimera,"* I said, using the only Greek I knew. "Hello." He smiled with rheumy eyes and motioned at me to join him for a drink. My feet were sore, Thera was hypnotic, the raki was inviting, and Artemis was delightful, so I did. Using sign language, I told him that I was planning to visit Akrotiri the next day.

My timing was lucky. Until recently Akrotiri had been off-limits, closed for eight years after a Spanish tourist died when part of the excavation's roof caved in. The site had just reopened, and now the smothered town that had produced so many timeless dolphin artworks was once again accessible to see. I had read about it endlessly, and I knew that any investigation of the Minoans' maritime obsessions needed to start in its abandoned streets.

What did I hope to find there? As the night rose up from the silence of the caldera, I thought about this. I suppose I wanted to know how a prosperous people had managed to live in unity with nature—for thousands of years. In the modern world, armed to the teeth with technological might, we have chosen another route. We believe in dominion: nature is ours to do with as we please. There is nothing we aren't willing to tamper with, even our own genetic code. Not long before I'd come to Greece, a paper was released estimating that half the world's animals have been exterminated in only two generations—and that was before you considered what has happened to plants and insects, coral reefs and rain forests and other ecosystems. We are causing extinctions, scientists say, at a thousand times the usual rate. We *know* we are mowing down life—we've made charts and graphs and we count everything on spreadsheets—but that understanding hasn't slowed us in the least. Harmony was a quaint idea for the rustics, our actions say, but we have more ambitious plans. And to put it plainly, our approach wasn't working so well. "We think we have understood everything," philosopher Thomas Berry wrote. "But we have not. We have *used* everything."

What was it about the Minoans that made them embrace nature—especially the ocean—rather than fearing it, exploiting it, or trying to conquer it, like other civilizations that followed them? Why were dolphins given such an eminent place in their world, along with other recurring creatures: birds, bees, bulls, snakes, lions, and octopi cropped up often in their art. Trees were another favorite subject, and so were flowers. And what about the Minoan fascination with spirals? They drew that symbol everywhere. What did these things mean to them—and what, after all these years, should they mean to anyone else?

≈

Akrotiri was a city of ash, layers and layers and layers of ash, and to make out the contours of its buildings and streets—still partially entombed, strewn with pottery, and braced with scaffolding—my eyes had to adjust to the monochrome view. There was an enveloping stillness that almost seemed to be breathing; an aura of sweetness tinged with lament. I was dumbstruck, and given the expressions on other people's faces as they entered, this was a typical response. The sight of a buried city is one you can never really prepare for, or easily make sense of. Which is where my guide, Lefteris Zorzos, came in.

Zorzos was the youngest archaeologist ever to work at Akrotiri, beginning his lifelong devotion to the site in 1999, at age sixteen. He was a dashing Greek, Athens-born and London-educated, a founding member of the Society for Aegean Prehistory, and someone who could elucidate the past with astounding detail. In his studies, Zorzos had identified thirty Minoan sites around Thera—and that was just for starters. "There's quite a bit underneath everything we're standing on," he said, in a refined accent with clipped corners. "When the eruption happened it wasn't just Akrotiri that was preserved—the entire island was preserved."

The day was as hot as lava, but under the climate-controlled roof the air was cool and dry. From the beginning, Marinatos had realized that Akrotiri yielded such an avalanche of intelligence, so many artifacts, that no one collection could ever hold them, and lobbied instead to turn the whole dig into a museum. Over time the Greek government, moving at glacial bureaucratic pace, had shelled out to protect the Minoan treasures, but the country's current economic crisis had halted any further work. "There hasn't been an excavation in several years," Zorzos said, in a rueful voice. "Very little research, very little conservation. So we are fighting on a local level to get things going." Often, he told me, he paid out of his own pocket for Akrotiri's essential expenses: "In 2000, we had a hundred archaeologists working here. Now we have just four or five guards. You can imagine the difference."

We stood near the entrance on an elevated planked floor that was level with the top stories of the buildings. From there, visitors could peer down on the city's layout as it was when the eruption struck. Only one structure had been fully uncovered; the rest were revealed only in sections. The site was expansive, but it likely represented only 3 percent of Akrotiri's full footprint. This had been a bustling place, home to thousands of people, conveniently situated by the sea like a Minoan San Francisco or Sydney. No one knew the extent of its boundaries: part of the settlement might lie at the bottom of the caldera now, or it might rest farther inland, still hidden under the ground.

Most of Akrotiri's buildings were private houses, Zorzos said, but a few had the grand appearance of public gathering spaces. "What we're seeing is the urban core," he pointed out. "And if you look below, you can see the sewer system running underneath the street level. It was a very, very advanced society."

The most electrifying find at Akrotiri was that the walls of every building had been painted with vibrant Minoan frescoes, and these artworks had survived under the ash. "This has never been seen before *anywhere*," Zorzos emphasized. Pottery of all shapes and sizes filled the houses, decorated with spirals and flowers and birds and dolphins; some buildings' floors had shimmered with crushed seashells. Akrotiri's post at the water's edge made it an obvious place for Minoan marine dreams to have blossomed. "They do have dolphins in many frescoes," Zorzos said, "and in a lot of the pottery. You also see them on seal stones. The animal was definitely significant to them."

We toured the settlement, Zorzos pointing out how the blanketing layers corresponded to the phases of the eruption. When Thera's volcano first cleared its throat, it spewed out a coarse pumice, followed by a torrential silvery hail. At some point, boulders had been ejected like cannonballs, crashing through walls, and they could still be seen, lying where they'd landed. Early in the destruction, the temperature inside the town hit 300 degrees Celsius. Then things got even hotter: "The second phase was four hundred degrees," Zorzos said. "So you're talking about complete annihilation."

He showed me the West House next, a structure that had contained so many ocean-themed artworks that researchers suspected it was the home of a high-ranking seaman, an admiral perhaps, or a ship captain. "This is probably one of the most important buildings in the Mediterranean," Zorzos said, with an air of nostalgia. In his first assignment at Akrotiri he had excavated a clay pipe on its ground floor, part of an efficient, anti-erosion drainage system. There was also a stone toilet that flushed, the earliest ever found.

The West House rose two stories high, with a wide stone staircase connecting the floors. Picture windows in its facade looked out on a triangular town square. The top floor had been wrapped in a 360-degree frieze of Thera's sailing fleet; this 39-foot-long, 17-inch-high painting was so intimately wrought it was as though the artist were documenting the scene like a newspaper reporter. The panels illustrated a red sand beach that still exists next to Akrotiri, with seven oceangoing ships and assorted smaller boats triumphantly circling the harbor. The city was drawn with precision too, thronged with people dressed for a party. (In frescoes, the Minoan women are always decked out in jewelry, wearing voluptuously flounced skirts. Above the waist they tended to go topless, their breasts showcased for admiration.) One nautical historian pointed out that in the fresco the Minoan ships appeared to have hydroplanes under their sterns, a feat of engineering that seemed impossibly futuristic for 1,600 BC. Personally, what I'd noted first were the festive pods of dolphins: they were everywhere, painted in cobalt, scarlet, rust, and ochre, vaulting over the ships, escorting the flotilla, and in some cases, even mingling among the crowds on shore.

Zorzos had saved one corner for last, a monumental, three-story building known as Xeste 3. It had been constructed with masonry blocks of volcanic stone, fitted together as seamlessly as matching puzzle pieces. "They had ingenious ways of keeping these tall buildings upright," Zorzos explained. "Every floor is built slightly further inward to make it stable. I've discussed it with architects—what you're seeing is very difficult to do even with cement today." Archaeologists believed that Xeste 3 had a sacred purpose; that it was a place with special intrigue. Considering what they found inside it, this was an excellent guess.

The lower two floors were a maze of rooms, many of them small and cloistered. At least one room was sunken, and probably used for ritual cleansing or bathing. But the building's most phenomenal feature, Zorzos stressed, was its artwork: in Xeste 3, not a single vertical surface had been left unadorned. These outlandishly fabulous frescoes, some crumbled into pieces by the eruption, had been carefully removed and transferred to a workshop on site, where they were restored for years before being loaned to museums. This process was closed to the public, but Zorzos had arranged for a private audience. In particular, he wanted to show me the paintings that had enveloped Xeste 3's top floor. "I will not say anymore," he declared, with a knowing grin. "Seeing them . . . well, I don't think you will believe what you see." He checked his watch and nodded. "So let's go. I think we are ready."

≈

Litsa was a pocket genie with a pixie haircut, maybe five feet tall and a hundred pounds and most of it was her smile. She leaned her whole body into the task of pulling the ten-foot-square fresco out on its rollers, digging in with her silver sneakers. I gasped when I saw the painting, and Zorzos beamed because he'd been expecting my reaction. Potnia flooded the room, casting magic on even the beige electrical sockets. The goddess was unspeakably lovely. She was seated in profile, leaning forward so you could see her cascading black ponytail, dotted with rubies, and beside her were the ornate, feathered wings of a griffin. She wore a necklace of dragonflies, and earrings like gold moons; crocuses, the emblem of spring, were embroidered on her flowing dress. Below her, appearing as smaller figures, a girl and a blue monkey made offerings. The fresco's colors—crimson, navy, gold, ivory, white, black—were as luminous as if they had been laid down yesterday. Maybe half the fragments had been refixed into place; this was a work in progress that, when finished, would reveal the pinnacle of Xeste 3. When the Minoans came to worship they had ascended to the top of this building, and this is whom they expected to find. This lady represented everything to them. Her calling card was benevolence: a love for all things clawed and finned and winged and flowering, the creatures that scampered and swam and flew and grew on this earth. She was Mother Nature, and never had I seen her in such corporeal glory. All I could say in the face of her image was, "Oh my God," which was, if you thought about it, pretty ironic.

"The level of detail is incredible," Zorzos said, in a hushed voice. "You can see it in her hands. Her fingernails are painted; her hair is very elaborate. Her dress itself is absolutely spectacular—there is an actual landscape on it: swallows, lilies, everything together. It has a lot of energy and power." Every piece we were seeing, he told me, had been painstakingly reassembled by Litsa and her team. "There are many more frescoes being worked on," he added.

The next panels Litsa pulled out were enormous, embellished spirals, huge blue whorls that had presided over Xeste 3's top floor, sitting above the goddess. "We haven't identified any windows in the building," Zorzos said, "so this may have been something that was completely closed off." Viewing them, I was lost for superlatives. It was easy to imagine falling on your knees in front of these frescoes, which was likely the point. The Minoans, it seemed, were determined to immerse themselves in beauty, sensuality, and wonder, a set of intentions that diverge radically from our own. That wasn't the only difference between us, either.

Scholars believe the Minoans' ubiquitous spirals represented a circular view of time, as opposed to the linear one we've adopted. In their outlook, birth, life, death, and rebirth existed in an unbroken continuum. The goddess's touchstones were the ever-regenerating moon, the constantly cycling seasons. To the Minoans, life was not a highway that dead-ended in a cliff: it wound on forever, and throughout all of it, nature supported them. In this context, the dolphins who popped up so buoyantly were the guides, the guardians, the intermediaries between this world and the underworld, and then back again. The word "dolphin" in Greek is *delphis,* which is, not coincidentally, almost identical to *delphys*—the womb. Out of the darkness below the surface comes the inevitable rebirth, with the dolphins as its midwives.

This may seem like a lot to infer from some pots and rubble, a fixation with sea creatures, a bellicose volcano, a drowned people—but to me, it all rang true. Anyone could see that even the humblest Minoan objects exuded a reverence for—and a profound connection to—the natural world. That connection had frayed now, but perhaps it wasn't completely severed. Maybe, with the help of some primordial memories, we could renew it. Craig Barnes, the author of a book called *In Search of the Lost Feminine,* summed this up hopefully: "The Minoans were not an anomaly or an aberration or a mystery of the deep past," he wrote, "so much as an early and vivid example of some innate quality of the human that does not ever go away."

Whatever name that quality had, whatever its etheric substance, it still infused these ruins. It seeped out of doors and wafted out of windows, and the frescoes positively oozed with it. I wondered if Zorzos, who had spent so much time among the Minoan ghosts, had picked up on this vibe. Walking around Akrotiri late at night, for instance, had he ever sensed its presence?

He answered emphatically. "Yes . . . but it's hard to describe."

"Is it a good feeling?" I asked, probing.

Zorzos smiled slowly. "Yes," he said. "Definitely, yes."

≈≈≈

The road was unremarkable, a typical thoroughfare running south and slightly uphill from Crete's dreary capital city, Heraklion. Anything formerly gorgeous or poetic along its route had long been paved over. There were markets and tavernas and T-shirt shops and car-repair joints, the usual commercial dross, nothing at all to indicate that if you followed this road, it would lead you directly to the dazzling zenith of the Minoan world: the Palace of Knossos.

In the late nineteenth century, archaeologists—alerted by farmers and shepherds who had been tripping over scraps of pottery and the tops of buried walls for years—had become interested in a promising stretch of Cretan fields, set among low-lying hills. There was something down there, they'd realized, something *big*. A wealthy Englishman named Sir Arthur Evans ultimately scooped them all, buying the land outright and embarking on a full excavation in 1900.

It was a heady time for hunters of lost civilizations. The German archaeologist Heinrich Schliemann had just unearthed Troy and Mycenae, validating Homer's accuracy and hauling up boatloads of gold; just about anybody willing to put in the elbow grease could find something primeval and priceless kicking around. When Evans first visited Crete, he was shocked to find the local women wearing necklaces of carved gemstones they'd collected in the fields. Walking around, Evans himself had more or less stubbed his toe on a clay tablet covered with inscriptions, markings that resembled nothing in linguistic history.

(This Minoan language is now known as Linear A, and to this day it remains undeciphered. Many more examples of it would be found on Crete and Thera, including the famous Phaistos Disc, a circle of fired clay about the size and shape of a mini-pizza. The disc was stamped 241 times with cryptic, pictorial symbols arranged in a spiral on both sides. Some of the symbols are familiar: a bird in flight, a woman in a long skirt, a man's head, an oar, a beehive, a leafy branch, a snake, the flower of life. Others were utterly arcane. Countless scholars have tackled the translation; none have succeeded. But there is one thing I can tell you for sure about the Phaistos Disc: the image of a dolphin appears on it five times.)

Having secured his site, Evans dug with gusto—and Knossos proved to be even more amazing than anyone had suspected. It was a vast complex of top-to-bottom architectural splendor, stuffed with the Minoans' signature artworks. One of his most celebrated finds was a suite of rooms known as the Queen's Megaron, the centerpiece of which was a panoramic fresco of dolphins, painted as though underwater. In Thera, I had noticed other dolphin artworks rendered from this perspective, including two altar tables that showed the animals hunting beak-down in the sea grass—a specific foraging technique they do use—and idly wondered how the Minoans would have known what the dolphins were doing below the surface. But then I read that Evans had turned up Minoan magnifying lenses and mirrors made of crystal. It was delicious to imagine that someday we might find a Minoan scuba mask on the Aegean seafloor—who would bet against this? Whatever their means, it was clear these dolphin artists had observed their subjects in action.

I'd been warned that Knossos was a mosh pit of sightseers, so I arrived early in the morning, but I still found myself queuing for a ticket behind busloads of people. It was an inferno of a day and tourists milled at the entrance, fanning themselves with guidebooks. Guides wearing earpieces held signs aloft and attempted to herd their charges;

vendors hawking Knossos posters and tote bags and fridge magnets swarmed outside the gates.

I decided to do the tour backward, reasoning that I'd be free of the crowds until at least the halfway point. Knossos was colossal. Its focal point was a central courtyard the size of a football field. This arena's purpose might have gone down as another puzzle, but the Minoans themselves had demonstrated it for us. Evans found an immense fresco that depicted a highly risky (and understandably defunct) sport. It involved a charging aurochs bull—a now-extinct species the size of a rhino, with cruel, impaling horns—along with three athletes, a man and two women. One woman faced the bull, grasping its horns, while the other woman tossed the man, handspring-style, onto the animal's back. It was a harrowing activity, but the scene had a gleeful quality: there was nothing gladiatorial about it. These people were *playing*, albeit in a don't-try-this-at-home sort of way. Bull leaping was clearly a beloved Minoan pastime—it appears on numerous artifacts.

Wandering through Knossos, I was staggered by the sheer volume of stone that had been used to construct the place. There were broad stone roads, stone staircases on the scale of the Mayan pyramids, facades made of limestone, floors and rooms lined with gypsum, a crystalline alabaster. Evans, searching for a way to describe what he found, had coined the term "palace," though he was far from certain the word summed up Knossos's role. It was a *hub*, that much was apparent, with multiple functions. The Minoans had gathered here for events and rituals and celebrations; they'd kept grain and olive oil and wine in an extensive web of storerooms. Using hydraulics, a bogglingly tough science, they'd built a system of clay pipes to divert and filter drinking water from a nearby spring. Artists had worked their alchemy on these grounds: metalsmithing, painting, stonecutting, and pottery studios had been excavated. And in a quiet, shaded room, not particularly big or imposing, someone had sat on a carved gypsum throne with a tall, scalloped back, flanked by a pair of sentinel griffins. Judging from the throne's size, shape, and décor, and the artwork and stonework that had been found along with it, researchers believed that someone had been a woman.

Around the corner from this elegant seat, I found the Queen's Megaron. The tour groups hadn't descended yet, so for a moment, anyway, I had the place to myself. In architectural terms, the word "megaron" means *great hall*, but to my eye this was more like a cozy hall. It contained a bathing area and adjoining lounges—boudoirs, perhaps. Evans had left several ceramic urns where he'd found them, each one whirling with spirals. There were nooks that might have been closets, a sunken basin, and another improbable flush toilet.

But mostly, there were dolphins. The fresco spanned the width of the main room, atop a double doorway. Five life-size dolphins swam in profile among schools of fish and bottom-dwelling sea urchins. The dolphins' eyes were painted exactly, their markings carefully and stylishly reproduced. Below them, spiral patterns linked up with a border of delicate rosettes. I leaned forward to examine the images better, feeling a deep sense of elation.

If Evans's notes are any indication, he had experienced the same thing. Though the fresco had been knocked off the wall and lay in a heap of pieces, when it was reconstructed even the man who'd peered into every miraculous corner of Knossos was impressed. "Large dolphins and numerous smaller fry were most naturalistically rendered," he wrote, marveling that "the spray and bubbles fly off at a tangent from the fins and tails, and give the whole a sense of motion that could not otherwise be attained . . ." Evans also raved about "the spirited character of the designs, the prevailing colors of the fish, blue of varying shades, black and yellow, the submarine rocks with their coralline attachments, and still more the manner of indicating the sea itself."

My reverie was ended by the arrival of an excited herd of tourists. Reluctantly I moved on, giving the fresco a final, backward glance. More than anything, I wanted to move into this megaron, to channel the Minoan high priestess or goddess or queen; to pour water into her alabaster bathtub and lay back and lounge here and feel what it was like to live in a society that took the ocean as its muse. I wished with every cell of my body that I could travel back in time and catch even the briefest glimpse of the person who had painted these dolphins. I wanted in on the secrets; I longed, as the Minoans did, to visit the depths and the heavens where only Potnia could take me. Though I was a card-carrying member of a culture convinced of its own linearity, I yearned to know life as a spiral.

≈≈≈

A chilly wind blew low across Elounda Bay, reminding everyone that summer was gone. In autumn, this fishing village on Crete's northeast coast became sleepy, settling in for the winter. The seaside cafés had put out a few lonely tables, most of which sat empty. A row of trawlers rested in their slips, tangles of nets and traps and gaffs ready on their decks. Some boats had dolphins painted on their bows, as if to summon good luck in finding the fish. These days in the Mediterranean, that wasn't such an easy proposition. The region's seas had been overfished to the point of collapse, the once-plentiful tuna and sea bass and grouper, the sharks and rays, all but gone. The dolphins had followed them, their populations crashing. Striped dolphins, common dolphins, and bottlenoses had been among the Mediterranean's liveliest and most visible inhabitants; now, it was rare to spot a single fin. Biologists were angling hard for the development of protected marine reserves, so the ecosystem could heal itself. Nature is stunningly resilient; if left alone for even a short time, the ocean rebounds fast. Marine reserves have flourished in other places, creating cause for optimism. But the time to act was yesterday—and so far, depressingly, that hadn't happened.

Across Elounda's town square, past the Dolphin taverna, the Dolphin grocery store, and the Dolphin hotel, I met a woman who was selling dolphin jewelry on the sidewalk. "There used to be many dolphins here," she told me, somewhat apologetically. "The Minoans, they loved the dolphins."

After visiting Knossos, I had left Heraklion and driven east. Elounda was an ideal base for the last leg of my trip: before I left Crete, I wanted to roam around the prefecture of Lasithi, a major Minoan stomping ground. Three more palaces—Malia, Gournia, and Zakros—had been found in this area, smaller, but perfectly situated on natural harbors. In these complexes, too, archaeologists had turned up the usual wonders: multistoried stone buildings, immaculate plumbing, spacious courtyards, sanctuaries with triple doorways, Linear A tablets, and a trove of marine-style artworks. At Zakros, they'd discovered a cistern that resembled a swimming pool, a smelting furnace with smartly designed ducts, and wine presses, along with some ten thousand precious objects: frescoes, cut-crystal vases, bronze and copper ingots, ivory tusks, ceramics, and a cache of five hundred clay seals carved with fantastic creatures, including women with eagles' heads and butterfly wings.

Amid all these riches, all this opulence, one very conspicuous thing was missing in the Minoan world: money. No coins, no currency, no gold standard—none of these had been found. Even the suggestion of money was absent. But the Minoans had traded, and they lacked for nothing, as their overflowing storerooms attested. How Minoan wealth was distributed, in any case, remains unknown. Considering that their artwork contains no images of transactions, it seems obvious that however they operated, the idea of individual profit—so urgent to our own civilization—was unimportant to them. More likely their goal was to enrich their whole society, and undeniably, what the Minoans appeared to value most were unquantifiable things like joy and freedom—and dolphins.

≈≈

I was relaxing on the pebbly beach at Elounda, gazing out at Spinalonga, a forlorn island about a mile offshore that had been used in the Byzantine era as a fortress, and then later became a leper colony. It was known, my guidebook said, for its "turbulent history of fierce battles and much human suffering." Earlier I'd considered taking a boat ride over there, but that excursion would have to wait because I had just read a paragraph that diverted all of my attention elsewhere. In my sheaf of Minoan research papers, Lasithi background, historical references for Crete, I had come across the following passage: "The picturesque fishing village of Elounda boasts the remains of the Minoan city of Olous. Folklore suggests that Olous may in fact be the lost city of Atlantis, and when the waters of the bay are calm, it is possible to snorkel over the walls to explore the sunken remains of this ancient site. Swimming here is good—however, watch out for sea urchins. Sadly, all that remains is a beautiful mosaic floor featuring dolphins frolicking."

I sat up in my lounge chair. This was the first I'd ever heard of Olous. Reading further, I learned that it had been a busy Minoan port. As many as thirty thousand people had once lived there; Olous had a close affiliation with Knossos. The city's ruins now lay in shallow water, only five miles from my hotel. After the Minoans quit Olous, others had moved in, and its later occupants had minted coins with Britomartis, the mermaid goddess, on one side, and a dolphin on the other. These coins, I read, were now on display at the Louvre.

Really? *A sunken Minoan dolphin city?* And it was right here? I gathered my things and bolted for my room. I didn't have a snorkel, but that wasn't going to stop me.

≈≈

The directions to Olous were tricky. I drove back through Elounda village, then hung a left at a cobblestone alley next to the Dolphin apartments. The alley led to a shoreline road that tracked along an isthmus, so low to the sea that a ripple would've swamped it. I knew I was on the right path when I passed the Britomartis Motel, a down-at-the-heels hostel the goddess would not have enjoyed. Five minutes later, I crossed a stone bridge of uncertain vintage—maybe prehistoric, maybe merely old—and then the road ended at the base of some undulating foothills, clumpy with brush. There wasn't much here: a few shanty houses, a shuttered taverna, feral cats. I pulled my car onto a patch of hardpan dirt, got out, and looked around. The air smelled briny and rich, as though it had been boiled down into a saltwater consommé. All I could hear was the slightest shush of the wind.

I set out for the mosaic first, picking my way across a rocky field. There were no signs or guideposts but there was a trampled footpath, so I followed it, disturbing a herd of goats and a jittery lizard with a mustard-colored head. I hiked past a derelict boat keeled over in the scrub grass, with a smattering of snail shells around it. The trail led to a wire fence that enclosed a space the size of a tennis court. Apparently a church had once stood here, but now only the floor was left—and it was made of dolphins.

Four of them swam across the ruin, escorted by little fish. Somebody must have spent years assembling them, setting thousands of dime-size black and white stones into contrasting patterns. The dolphins were black, with exaggerated white eyes and round black pupils; their beaks were white, with a black outline. Their bodies were curved, as if to suggest liquid movement, which couldn't have been easy in this medium. Around them, the floor exploded into checkerboards, lines, triangles, arches, flowers, shapes that were meant as waves. The center of the floor had been scoured by time, but the dolphins were immaculate. They wouldn't last forever, of course, given that they were totally exposed to the weather.

I knew that the artist had not been Minoan; this was a later offering, from sometime closer to Christianity. And as charming as the mosaic was, it lacked the Minoan touch. These dolphins were a bit jagged, crude almost, when compared to the frescoes—and also, the spirals were gone. Although we like to equate the passage of time to the forward march of progress, after the Minoans left, a dark age reigned. Over the centuries Crete had been conquered and pillaged and sacked and invaded, over and over—by the Mycenaeans, the Dorians, some aggressive barbarians known as the Sea Peoples, the Romans, the Arabs, the Venetians, the Turks, and even Nazi Germany. This place had been pummeled, and yet somehow its dolphin iconography remained.

I lingered for a while and then turned back, anxious to find the sunken Olous. The light had ripened, the shadows had lengthened, and as I walked I began to notice the silhouettes of structures. I stopped in my tracks. Earlier, this place had seemed desolate; now I saw that every hillside was crosshatched with stone walls. This *was* a city, but rather than being tended, excavated, and protected like Knossos or Zakros or Akrotiri, it was merging back into the earth.

I followed a wall that led straight into the water, and then stripped down to my bathing suit. Even from above I could tell the visibility was stellar, and suddenly I could see what lay below the surface. There were ruins down there, vague but unmistakable. As I dropped my clothes on the shore, I realized that shards of pottery were everywhere underfoot. Some of them were even painted. I stepped carefully across the rocks and slipped in.

The sea was aquamarine, and as clear as glass. Opalescent fish darted around me as I stroked into the bay, above a rampart that gradually dropped off. I dove, and saw that on the bottom, algae sprouted between the cracks in its stone foundations. Hours passed like minutes as I traced Olous in its aquatic afterlife. Sea urchins guarded doorways; a rainbow-bellied eel flashed in the late illumination of the day.

Atlantis this was not, but Olous still made my spirit soar. I swam over to a staircase that led into the depths, and I felt the Minoans beckoning me.

What—and who—was down there? What mysteries lived beneath the ocean's blue skin? Only the dolphins knew. If there was anything I had learned it was that they, not we, were the masters of their element. Their voices were not ours, their language was unknown to us, but if we listened we could hear their song. It was an ageless, rapturous melody, only faintly heard but somehow known by heart. It echoed through the waters; it rang across the shoals of life.

What if nature spoke to us in music, and the dolphins were her chorus? What if we stopped talking, and joined their harmony?

What if the world was singing to us all the time?

ACKNOWLEDGMENTS

My thanks to the sea of dolphin lovers who helped me during every phase of this book: there are so many of them, and they all care deeply about the ocean and the magnificent creatures who live in it.

In the science world, I am especially indebted to Robin Baird, a brilliant biologist who was always generous with his time and knowledge. I also thank his colleagues Daniel Webster and Brenda Rone, and the other scientists at the Cascadia Research Collective. I hope readers will come away with a true understanding of the dedication and skill it takes to study cetaceans in the wild, as exemplified by this group.

I also owe major thanks to Lori Marino, whose work is as inspiring as it is informative. As I wrote, I asked Lori countless questions, all of which she answered with her signature combination of razor-sharp intellect, great humor, big heart, and patience. I look forward to following her next groundbreaking project, the Kimmela Center for Animal Advocacy.

During my reporting I was constantly aware of the efforts of individuals and environmental groups to fight necessary battles on behalf of dolphins and whales. In this realm, I am exceptionally grateful to Ric O'Barry. From the moment I was introduced to him in the movie *The Cove*, Ric gained my admiration and respect, feelings that only grew stronger when I got to know him personally. His devotion to dolphins is legendary, and he has spurred an entire generation of activists. I also wish to thank his son, Lincoln O'Barry.

I met many other fearless and wonderful dolphin advocates while tracing Ric O'Barry's travels. My thanks go to all of them, and especially: Tim Burns, Carrie Shadley Burns, Masako Maxwell, Terran Vincent Baylor, Arielle Peri, Victoria Hawley, Becca Jurczak, Melissa Thompson Esaia, Vickie Collins, Jeremy Raphael, Veronica Artieda, Daniela Moreno, Russ Ligtas, Sakura Paia, Yaz Riddler, Brittany Clack, Kiki Tanaka, Jess Chan, and the awesome Cara Sands. On the subject of Marineland Canada I am indebted to the investigative work of *Toronto Star* reporters Linda Diebel and Liam Casey, whose articles did more than educate readers: they prompted government action. I thank all of the Marineland whistleblowers, individuals who stood up for the animals at heavy personal cost. Likewise, I thank the Cove Monitors, everyone who has gone to Taiji to work for a better future there—for dolphins and people. I can attest that it is not an easy place.

At Earth Island Institute, Mark Berman and Mark Palmer provided much support and guidance, both in Taiji and for my trip to the Solomon Islands. The ocean has formidable friends in these two men, and in their colleague, Lawrence Makili. In Honiara I was also grateful for the acquaintance, and the help, of Anthony Turner. In advance of my visit, Byron Washom and Kelly Siman gave me much needed advice about the islands. Chris Porter spoke to me candidly about his experiences at Gavutu and their aftermath.

Further thanks go to Frances Beinecke, Joel Reynolds, and Michael Jasny at the Natural Resources Defense Council, and David Henkin at EarthJustice. Both groups work tirelessly—and effectively—to protect the natural world, in times when the stakes have never been higher. Rising noise and toxic pollution in the oceans are vexing problems, and I greatly appreciate the time they spent explaining these issues to me. Likewise, I am grateful to Mati Waiya, Luhui'Isha Waiya, Jason Weiner, and everyone involved with Wishtoyo Chumash Village, a beautiful model of environmental stewardship.

When I returned to Hawaii to continue my dolphin investigations, I was fortunate to have met Joan Ocean, Jean-Luc Bozzoli, and their friends. The time I spent with them was amazing, and I am thankful for their contributions to this book, and for their ongoing role in engendering a heartfelt connection to cetaceans. Not long after I finished my reporting I had the sublime experience of being in the water with Joan and a pair of gentle, curious humpback whales. I watched from six feet away as one of the whales beelined toward her and basically kissed her underwater, twirling his forty-ton body in delight. If you proceed with love, Joan showed me, you will find it all around you.

No place I traveled to for *Voices in the Ocean* was more awe-inspiring than Thera. The Minoans may be gone but their stunning artworks remain, so many of them heralding dolphins. For direction about that time and that region, I am indebted to marine scientists Katy Croff Bell and Evi Nomikou, who provided background information and the introduction to archaeologist Lefteris Zorzos. I am grateful to Lefteris for his kindness and his expertise, and I urge interested readers to delve further into the excavations at Akrotiri. The ideal base for your explorations is the Voreina Gallery Suites, a jewel of a hotel run by Lefteris, only ten minutes from the site.

As always, I owe much to my editor, Bill Thomas, who somehow manages to make the years spent working on a book feel like fun. He guided this project from its beginning, and his expertise is present on every page. My gratitude to him is immeasurable. Likewise, I thank his colleagues at Knopf Doubleday for their contributions: Maria Carella, Rose Courteau, Melissa Danaczko, Todd Doughty, John Fontana, Suzanne Herz, Lauren Hesse, Kathy Hourigan, Lorraine Hyland, Lawrence Krauser, Nora Reichard, and Anke Steinecke. At Random House of Canada, I thank Kristin Cochrane, Amy Black, Josh Glover, Brian Rogers, and the rest of their team.

My agent, Eric Simonoff, is a true partner and a constant source of wisdom. His advice and insight are unfailingly smart, and always delivered with warmth and great humor. The same is true of Naomi Barr, researcher extraordinaire. Her dedication, intelligence, and curiosity are evident throughout this volume.

Tim Carvell, Terry McDonell, and Sara Corbett read early versions of my manuscript and offered their usual perfect counsel. In New York, I also owe thanks to Ellen Levine, David Carey, Eliot Kaplan, Lucy Kaylin, Adam Glassman, Karla Gonzalez, and David Granger at the Hearst Corporation. As always, I thank my family: my brother, Bob Casey, my sister-in-law, Pamela Manning Casey, and my mother, Angela Casey, whose love of animals has inspired me throughout my life.

In Hawaii, I am blessed to be surrounded by friends who embody the true spirit of aloha: Donna Palomino Shearer, Don Shearer, Deborah Caulfield Rybak, Michael Rybak, Judie Vivian, Rob Vivian, Devri Schultz, Teddy Casil, Skeeter Tichnor, Suryamayi Aswini, Rich Landry, Linda

Sparks, Karen Bouris, Nancy Meola, Paul Atkins, Gracie Atkins, Gabrielle Reece, Laird Hamilton, Shep Gordon, Paula Merwin, and William Merwin. On the mainland, I thank Kelly Meyer, Ron Meyer, Ann Moss, Jerry Moss, Andy Astrachan, Jane Kachmer, Cristina Carlino, Caroline Myss, Gayle King, and Oprah Winfrey for their kindness and support.

Finally, thanks to my nearest, dearest, ridiculously magical ohana: Martha Beck, Karen Gerdes, Adam Beck, Boyd Varty, Koelle Simpson, Travis Stock, Bob Dandrew, Elizabeth Lindsey, Maria Moyer, Shaun Simmons, and above all, Rennio Maifredi.

NOTES

PROLOGUE: HONOLUA

6   *"Despite having sustained":* Michael Zasloff, "Observations on the Remarkable (and Mysterious) Wound-Healing Process of the Bottlenose Dolphin," *Journal of Investigative Dermatology* 131 (July 2011): 2503–5. Also: Maureen Langlois, "Shark Bites No Match for Dolphins' Powers of Healing," NPR interview, July 25, 2011.

7   *"It's like dolphins and whales":* Hal Whitehead, as quoted in Charles Siebert, "Watching Whales Watch Us," *The New York Times Magazine,* July 12, 2009.

8   *"other utterances":* Sam Ridgway, Donald Carder, Michelle Jeffries, and Mark Todd, "Spontaneous Human Speech Mimicry by a Cetacean," *Current Biology* 22 (October 2012): R860–R861.

9   *These adaptations aligned them more with humans:* Michael R. McGowen, Lawrence I. Grossman, Derek E. Wildman, "Dolphin Genome Provides Evidence for Adaptive Evolution of Nervous System Genes and a Molecular Rate Slowdown," *Proceedings of the Royal Society B Biological Sciences* 279, no. 1743 (June 27, 2012): 3643–51.

10   *"The future for dolphins":* "Future Bleak for Dolphins and Porpoises," *New Scientist,* December 3, 2005, 4.

11   *"Do [dolphins] have":* Rachel Smokler, *To Touch a Wild Dolphin: A Journey of Discovery with the Sea's Most Intelligent Creatures* (New York: Anchor Books, 2001), 12.

12   *"Slowly we gave up the attempt":* Loren Eiseley, *The Star Thrower* (New York: Harcourt Brace & Company, 1978), 37.

CHAPTER 1: THE MEANING OF WATER

22   *"the most mysterious of fauna"*: Kenneth S. Norris, "Looking at Wild Dolphin Schools," *Dolphin Societies: Discoveries and Puzzles* (Berkeley and Los Angeles: University of California Press, 1991), 7.

23   *"The Kealake'akua Bay topography was perfect"*: Ibid., 11.

25   *I had read about the shy nature:* Recently scientists have warned that the presence of swimmers and boats might be affecting the spinners' ability to rest in Hawaii's near shore bays. They have recommended long-term studies to determine whether constant human interaction is detrimental to the dolphins. In my experience the captains were careful, and conscious of the spinners, and they stressed the importance of not approaching the animals too closely, or in any kind of aggressive manner. During my swims in Kona, in fact, it was the dolphins who ventured closer to the people, rather than vice versa. Yet it is possible that even the most respectful snorkelers, if there are enough of them, might be disturbing the animals' regular patterns.

29   *"I view the school as the matrix"*: Kenneth S. Norris, letter to Dr. John Lilly, June 22, 1976.

CHAPTER 2: BABIES IN THE UNIVERSE

34   *"We discussed the possibilities"*: John C. Lilly, *Man and Dolphin* (New York: Doubleday & Company, 1961), 40–44.

35   *"There was a powerful"*: Ibid., 45–46.

35   *"I wondered how"*: Ibid., 43.

37   *"We put him back"*: Ibid., 55.

37   *"We were all shocked"*: Ibid., 56.

39   *"I was stimulated"*: Ibid., 61.

39   *"through small needles"*: Ibid., 64.

40   *"Whistles, buzzings"*: Ibid., 75–76.

40   *"I had the rather uneasy"*: Ibid., 76.

40   *"I suddenly realized"*: Ibid., 84.

41   *"We began to have feelings"*: John Lilly, MD, PhD, "Productive and Creative Research with Man and Dolphin." Paper presented at the Fifth Annual Lasker Lecture, Chicago, April 1962.

42   *"I visualize a project"*: John C. Lilly, *The Mind of the Dolphin* (New York: Avon Books, 1967), 66.

43   *"In psychological warfare"*: Lilly, *Man and Dolphin*, 220.

43   *"I started from scratch"*: Ibid., 136.

44   *"Within the next decade"*: Ibid., 11.

45   *"Scientifically unsound and naïve"*: Margaret C. Tavolga and William N. Tavolga, *Natural History* 71 (1962): 5–7.

45   *"Borderline irresponsible . . ."*: E. O. Wilson review of *Man and Dolphin and*

Mind of the Dolphin, quoted in *Cetacean Societies: Field Studies of Dolphins and Whales*, ed. Janet Mann, Richard C. Connor, Peter L. Tyack, and Hal Whitehead (Chicago: University of Chicago Press, 2000), 28–29.

45 *"I read practically everything"*: John C. Lilly, *The Center of the Cyclone: An Autobiography of Inner Space* (New York: Bantam Books, 1972), 2.

46 *"The first few nights"*: Lilly, *The Mind of the Dolphin*, 231.

46 *"naughty dolphin"*: Ibid., 231–39.

46 *"My bed now has"*: Ibid., 238–42.

47 *"He does not go away"*: Ibid., 259.

47 *"This is obviously a sexy"*: Ibid., 256.

48 *"Even if it revives"*: Stanford University Library, Lilly Papers, Call No. MO786, Box 11, manuscript titled, "The Cetacean Brain," by John C. Lilly, p. 4. Quoted in *Oceans* magazine, May 31, 1977.

48 *"Instantly I knew her"*: Lilly, *The Center of the Cyclone*, 228.

49 *"I think many"*: Stanford University Library, Lilly Papers, Call No. MO786, Box 11, correspondence between John Lilly and Ken Norris, November 20, 1980.

49 *"There are times"*: Lilly, *The Mind of the Dolphin*, 12.

49 *"Your letter is a breath"*: Stanford University Library, Lilly Papers, Call No. MO786, Box 11, correspondence between John Lilly and Ken Norris, December 30, 1980.

50 *"extraterrestrial who has come"*: Francis Jeffrey and John C. Lilly, *John Lilly So Far* (Los Angeles: Jeffrey P. Tarcher, 1990), 211.

50 *"My work now"*: Rex Weyler, *Song of the Whale* (Garden City, NY: Anchor Press), 52.

51 *"For $7,000/month"*: Stanford University Library, Lilly Papers, Call No. MO786, Box 11, Project Janus folder.

52 *There was a freewheeling quality*: Stanford University Library, Lilly Papers, Call No. MO786, Box 11, Project Janus folder.

53 *"Mankind likes the hero"*: Stanford University Library, Lilly Papers, Call No. MO786, Box 11, transcripts from Jenny O'Connor channeling the Nine, dated January 16, 1981, and February 3, 1981.

53 *"If I were from an older"*: Lilly, *The Mind of the Dolphin*, 62.

54 *"The whole philosophy"*: Lilly, *Man and Dolphin*, 55.

55 *"A certain willingness to face"*: John Lilly, "A Feeling of Weirdness," in *Mind in the Waters: A Book to Celebrate the Consciousness of Whales and Dolphins*, ed. Joan McIntyre (New York: Charles Scribner's Sons, 1974), 77.

CHAPTER 3: CLAPPY, CLAPPY!

57 *You would have to decide:* These prices and options were offered in February 2012, when I visited the park.

58 *Coincidentally, its name:* There is no relationship between these two marine parks.

58   *"the improper burial"*: Editorial, "Ocean World Getting Off Too Easy," *South Florida Sun Sentinel,* June 5, 1992.

59   *"It was an accident"*: Alan Cherry, "Marine Park Under Siege, Hires PR Firm," *South Florida Sun Sentinel,* November 25, 1990.

59   *Orcas and beluga whales are popular attractions:* One female orca from the Pacific Northwest's J-Pod is thought to be 103 years old. Female orcas have an average life expectancy of fifty years, with individuals living into their eighties and nineties and beyond; male orcas' average life expectancy is thirty years, with individuals living into their fifties and sixties and beyond. These are not the kinds of numbers typically seen among orcas in captivity. In a detailed analysis of marine mammal mortality rates, the World Society for the Protection of Animals and the Humane Society of the United States reported that "the overall mortality rate of captive orcas is at least two and a half times as high as that of wild orcas, and age- and sex-specific annual mortality rates range from two to six times as high" ("The Case Against Marine Mammals in Captivity," http://www.humanesociety.org/assets/pdfs/marine_mammals /case_against_marine_captivity.pdf). "Killer whales in captivity live into their teens if they're lucky, into their 20s if they're really lucky and maybe into their 30s if they're amazingly lucky," Naomi Rose, PhD, senior scientist for the Humane Society, told reporter Jeffrey Wright of the *San Antonio Current* ("So Wrong, But Thanks for All the Fish," April 14, 2010). Belugas tell a similar story. In the wild, belugas have life expectancies comparable to orcas'; in captivity, 50 percent of belugas die by age eight.

60   *History has shown:* Bottlenose longevity is one of the most contentious issues in cetacean captivity. In recent decades, husbandry has improved at many marine parks, but it is clear that captive dolphins suffer from numerous stressors that adversely affect their health. In the wild, a healthy bottlenose might live for fifty or more years; females as old as forty-five have given birth. To date, only one captive female bottlenose has lived to fifty. In 2004, the *South Florida Sun Sentinel* ran an investigative series about captive marine mammals in the United States; the paper wrote: "Over the past 30 years, according to federal records, more than 3,850 sea lions, seals, dolphins and whales have died while in the 'care' of humans. Of those, about one-fourth perished before the age of 1. Half died by the age of 7." Currently, approximately 2,400 cetaceans live in captivity in sixty-three countries, including 52 orcas, 227 beluga whales, and some 2,100 bottlenose dolphins. The highest concentrations of animals are in Japan, China, the United States, and Mexico (editorial, "Industry Must Improve Care," *South Florida Sun Sentinel,* May 23, 2004).

      For an excellent overview of related issues you can download "The Case Against Marine Mammals in Captivity," a report from the Humane Society of the United States and the World Society for the Protection of Animals: http://www.humanesociety.org/assets/pdfs/marine_mammals/case_against _marine_captivity.pdf.

62   *"exchange and disseminate"*: International Marine Animal Trainers' Associa-

tion, "IMATA's Mission Statement," http://www.imata.org/mission_values (accessed June 2014).

64 *"In August last"*: *Captive Belugas: A Historical Record and Inventory*, p. 2. http://www.ceta-base.com/library/cetabasedocs/captivebelugas_august2010.pdf (published August 24, 2010; accessed August 2014).

64 *"fearful death"*: "Disastrous Fire," *The New York Times*, July 14, 1865.

66 *"the peevish whining"*: Kritzler, "Observations of the Pilot Whale in Captivity," *Journal of Mammology* 33, no. 3 (August 1952): 321–34.

67 *"The dolphin does amazing things"*: Richard E. Shepard, "Flipper, the Educated Dolphin, Cavorts in a Seascape Drama," *The New York Times*, September 19, 1963.

67 *"Growing attendance"*: Tess Stynes and Michael Calia, "Seaworld Profit Up 30%," *The Wall Street Journal*, November 13, 2013.

67 *Jim Atchison:* In the wake of falling attendance and a plummeting stock price, Jim Atchison stepped down from his post as SeaWorld's CEO in December 2014. He was replaced by Joel Manby, who took over in April 2015.

68 *After a sixteen-hour:* Later it was discovered that, along with being subjected to the rave, the dolphins had been fed buprenorphine, a heroin substitute. In 2012, the Swiss government voted to end all live dolphin imports into the country, which resulted in Connyland closing its dolphin show in 2013.

68 *"a perversion"*: William Johnson, *The Rose-Tinted Menagerie* (Easthaven, CT: Inland Book Co., 1990), 220.

68 *They suspect that:* William A. Walker and James M. Coe, "Survey of Marine Debris Ingestion by Odontocete Cetaceans," in R. S. Shomura and M. L. Godfrey, eds., *Proceedings of the Second International Conference on Marine Debris* (Honolulu, Hawaii, April 1989), 747–50.

70 *Scientist Jason Bruck:* Jason N. Bruck, "Decades-Long Social Memory in Bottlenose Dolphins," *Proceedings of the Royal Society B: Biological Sciences* 280 (October 7, 2013).

70 *"Be wild, be free"*: Fox Butterfield, "Claiming Harassment, Aquarium Sues 3 Animal Rights Groups," *The New York Times*, October 1, 1991.

70 *No wonder marine parks:* Hormones, antifungals, and antipsychotics are also used extensively.

71 *When it comes to partners:* A note about the dolphin libido: dolphins are enthusiastic and aggressive about sex, and promiscuous, too. Scientists believe there are social reasons for this—bonding, or to establish a hierarchy—as well as the obvious ones: procreation and plain old enjoyment. Among males, displaying an erection is often a sign of dominance (though sometimes it might be a game, a practice session, or an invitation). Homosexual dolphin relations are common. So is masturbation: both male and female dolphins have been observed rubbing themselves against soft sand on the seafloor. Favorite dolphin pastimes include buzzing one another's genitals with their sonar, presumably a highly pleasurable sensation, and engaging in plenty of brushing and touching. There is a courtship element to their pairings, but sometimes dolphin sex can

get rough. Biologist Richard Connor, who has studied a bottlenose society in Shark Bay, Australia, for thirty years, discovered that the males travel in roving bands, forming alliances to corral females, sometimes against the females' will.

71 *"This has to be"*: Steve Hearn, "Trainers Forum," *Soundings* Q4, 2008. http://www.mydigitalpublication.com/article/Trainer's_Forum/69098/8105/article.html (accessed in February 2014).

71 *"internal strife"*: Elizabeth Batt, "Taiji Dolphin Injured at Dalian Laohutan Ocean Park in China?" *Digital Journal*, March 18, 2012. http://www.digitaljournal.com/article/321391 (accessed April 13, 2015).

72 · *"normal, socially induced"*: "Performing Whale Dies in Collision with Another," *The New York Times*, August 23, 1989.

72 *"It should be noted"*: Naomi A. Rose, E. C. M. Parsons, and Richard Farinato, "The Case Against Marine Mammals in Captivity," report from the Humane Society of the United States and the World Society for the Protection of Animals, p. 31: http://www.humanesociety.org/assets/pdfs/marine_mammals/case_against_marine_captivity.pdf.

72 *52 percent:* Tania D. Hunt, Michael H. Ziccardi, Frances M. D. Gulland, Pamela K. Yochem, David W. Hird, Teresa Rowles, and Jonna A. K. Mazet, "Health Risks for Marine Mammal Workers," *Diseases of Aquatic Organisms* 81 (2008): 81–92.

72 *"Trainer R first entered"*: G. A. Shurepova, "Aggressive Behavior of Captive Tursiops Truncatus," in K. K. Chapskii and V. E. Sokolov, eds., *Morphology and Ecology of Marine Mammals: Seals, Dolphins, Porpoises* (New York: John Wiley & Sons, 1971), 150–53.

73 *"aggressive manifestations"*: Bruce Stephens, as quoted in Erich Hoyt, "The Performing Orca: Why the Show Must Stop" (Bath, U.K.: Whale and Dolphin Conservation Society, 1992), 31.

73 *While other SeaWorld employees:* United States of America Occupational Safety and Health Review Commission, *Secretary of Labor, Complainant v. SeaWorld of Florida*, LLC OSHRC Docket No. 10-1705, June 11, 2011: 1–47. This tragedy has also been chronicled extensively by journalists David Kirby (in his book *Death at SeaWorld: Shamu and the Dark Side of Killer Whales in Captivity*, published in 2012 by St. Martin's Press) and Tim Zimmermann (in articles that appeared in *Outside* magazine, available on his Web site: www.timzimmermann.com).

74 *Daniel Dukes:* The official cause of Dukes's death was found to be hypothermia leading to drowning, but the coroner's report made it clear that while he was still alive, Dukes had been thoroughly mauled by Tilikum.

75 *his semen used:* Tilikum is SeaWorld's most prolific orca stud, having sired at least twenty-one calves, approximately eleven of whom remain alive as of this writing. It is estimated that 54 percent of the whales in SeaWorld's collection carry Tilikum's genes.

CHAPTER 4: THE FRIENDLIES

81 *"I still think and dream":* Maddalena Bearzi and Craig B. Stanford, *Beautiful Minds: The Parallel Lives of Great Apes and Dolphins* (Cambridge, MA, and London: Harvard University Press, 2008), 24–27.

81 *When surfer Todd Endris:* Marcus Sanders (as told to), *Norcal Shark Attack Update: Firsthand Account of Monterey Bay Surfer Badly Bitten at Marina State Beach,* SurfNews, Surfline.com, August 29, 2007. http://www.surfline.com /surf-news/firsthand-account-of-monterey-bay-surfer-badly-bitten-at-marina -state-beach-norcal-shark-attack-update_10788/ (accessed July 2014).

82 *Their consideration of us:* Bonnie J. Holmes and David T. Neil, "Gift Giving by Wild Bottlenose Dolphins (Tursiops sp.) to Humans at a Wild Dolphin Provisioning Program, Tangalooma, Australia," *Antrozoos* 25 (2012): 397–413.

85 *"He swam alongside":* Anthony Alpers, "The Story of Pelorus Jack," in Eleanore Devine and Martha Clark, eds., *The Dolphin Smile: 29 Centuries of Dolphin Lore* (New York: MacMillan Company, 1967), 200.

86 *In a 1991 documentary: The Dolphin's Gift,* directed and written by Kim Kindersley, Zari Productions, London (video), 1991.

89 *"You could sense":* Ibid.

93 *When two drunken men:* Marcos César de Oliveira Santos, "Lone Sociable Bottlenose Dolphin in Brazil: Human Fatality and Management," *Marine Mammal Science* 13 (April 1997): 355–56.

93 *"Slidell Memorial Hospital's press secretary":* David Freese, "Slidell Teen Was Third in a Week Bitten by Dolphin," *St. Tammany News,* New Orleans, LA, May 23, 2012.

94 *They worried aloud:* Alas, The Dolphin's story does not have a happy ending. After heavy rains diluted the canal's salinity, Slidell residents observed that the bottlenose had developed skin lesions and lost his characteristic swagger. He died in June 2014, his body found washed up onshore. "It was a sad day for the entire neighborhood," one resident told a local newspaper reporter. "We all felt like we had a pet dolphin. He will be missed."

94 *A 2008 global census:* Lissa Goodwin and Margaux Dodds, "Lone Rangers: A Report on Solitary Dolphins and Whales Including Recommendations for Their Protection," Marine Connection (London, England), March 2008.

95 *"with increasing frequency":* Kathleen M. Dudzinski and Toni Frohoff, *Dolphin Mysteries* (New Haven, CT, and London: Yale University Press, 2008), 136.

CHAPTER 5: WELCOME TO TAIJI

100 *a watchdog initiative started in 1990:* In the past, the tuna fishermen's preferred method for locating their quarry had been to track rowdy pods of feeding dolphins—who are, of course, the real tuna-finding experts—and then winch up every last creature using giant purse seine nets, later discarding the dolphins

and any other animal that wasn't a tuna. This onslaught, which began in the fifties, had killed untold millions of dolphins, sharks, and other fish. Now, any tuna company found setting its nets around dolphins is flagged, publicly identified, and prevented from displaying the Dolphin Safe insignia on its cans.

100 *The world had acted:* Though commercial whaling was banned by the International Whaling Commission in 1986, Japan, Norway, and Iceland have continued to hunt minke, sei (endangered), fin, humpback, Bryde's, and sperm whales. Norway and Iceland simply ignored the ban; Japan found a loophole and exploited it. Under the guise of "scientific research," its whalers continued to kill roughly a thousand whales each year, mainly in the Antarctic and the Northern Pacific. In 2014, noting that Japan had not produced any scientific papers or research that would have justified the death of even a single whale, the UN's International Court of Justice ruled the country's actions illegal. Japanese government officials expressed "deep disappointment" with the court's decision, and told reporters they were determined to continue whaling. Defiant, the men encouraged photographers to take pictures as they ate from a lavish whale meat buffet.

101 *Scientists now know:* Joe Roman, James A. Estes, Lyne Morissette, Craig Smith, Daniel Costa, James McCarthy, J. B. Nation, Stephen Nicol, Andrew Pershing, and Victor Smetacek, "Whales as Marine Ecosystem Engineers," *Frontiers in Ecology and the Environment* 12 (September 2014): 377–85. http://www.esajournals.org/doi/abs/10.1890/130220.

104 *"this killing method":* Andrew Butterworth, Philippe Brakes, Courtney S. Vail, and Diana Reiss, "A Veterinary and Behavioral Analysis of Dolphin Killing Methods Currently Used in the 'Drive Hunt' in Taiji, Japan," *Journal of Applied Animal Welfare Science* 16:2 (April 2013): 184–204.

108 *"This is a small town":* Martin Fackler, "Mercury Taint Divides a Japanese Whaling Town," *The New York Times,* February 21, 2008.

108 *"But they do it":* Mark Willacy, "Unlikely Allies Attempt to Stop Dolphin Killing," Australian Broadcasting Corporation, February 2, 2012. http://www.abc.net.au/7.30/content/2012/s3421925.htm.

108 *Though all dolphin meat:* Andy Coghlan, "It's Madness," *The New Scientist,* June 8, 2002: 17.

108 *other than advising:* Andy Coghlan, "Shops in Japan are Selling Mercury-Riddled Dolphin as Whalemeat," *The New Scientist,* June 14, 2003: 7.

108 *"There is a real danger":* Fackler, "Mercury Taint."

111 *In 2012, for instance, marine:* "Drive Fisheries: Capture Results and Information," http://www.ceta-base.com/drivefisheries.html#20122013. These statistics are also compiled annually in an official document titled "Japan. Progress Report on Small Cetacean Research."

115 *"The plaintiff, Ocean World":* In the Circuit Court of the Seventeenth Judicial Circuit in and for Broward County, Florida. Case No. 11-17871, *Ocean World, S.A., Plaintiff vs. Earth Island Institute, Inc., and Richard O'Barry, Defendants.* Page 5, Section 28, of General Allegations.

115 *still no end in sight:* By January 2014, all four suits brought against O'Barry (and others) by Ocean World had been either dismissed or settled, some confidentially and some not. Details are not available for the confidential settlements, but in the others, the settlements were made for nominal amounts. (Currently, a single related lawsuit remains in progress and it's a brand-new one: Ocean World is now suing its former lawyers, the ones who represented it throughout all of these cases.) Reading through reams of court transcripts, the circuitousness of which would have impressed Franz Kafka, I felt a deep sense of admiration for O'Barry's lawyers, Deanna Shullman and Rachel Fugate from the Lake Worth, Florida, firm of Thomas & LoCicero. Their years of work wrangling with these matters—and thus defending the First Amendment—much of it done pro bono, struck me as nothing short of heroic.

CHAPTER 6: A SENSE OF SELF

130 *"While they don't build rockets":* Bruce Dorminey, *Researchers Closer to Decoding Dolphin Speak,* Forbes.com, http://www.forbes.com/sites/bruce dorminey/2012/10/18/dolphin-speak-bustin-the-code-on-flippers-rhymes/ (October 18, 2012).

131 *But what happened during:* Lori Marino, Mark D. Uhen, Nicholas D. Pyenson, and Bruno Frohlich, "Reconstructing Cetacean Brain Evolution Using Computed Tomography," *The Anatomical Record* 272B (2003): 107–17.

131 *In 2000, she and another:* Diana Reiss and Lori Marino, "Mirror Self-Recognition in the Bottlenose Dolphin: A Case of Cognitive Convergence," Proceedings of the National Academy of Sciences for the United States of America 98, no. 10 (May 2001): 5937–42.

133 *Marino examined:* Statement of Lori Marino, PhD, Neuroscience and Behavioral Biology Program, Emory University, Atlanta, Georgia, to the House Committee on Natural Resources Subcommittee on Insular Affairs, Oceans and Wildlife, regarding educational aspects of public display of marine mammals, April 27, 2010.

133 *down to the lowly earthworm:* For an eye-opening essay on the consciousness of earthworms, please see: Eileen Crist, "The Inner Life of Earthworms: Darwin's Argument and Its Implications," in *The Cognitive Animal: Empirical and Theoretical Perspectives on Animal Cognition,* eds. Marc Bekoff, Colin Allen, and Gordon M. Burghardt (Cambridge, MA: MIT Press, 2002), 3–8.

133 *"The body of scientific evidence":* George Dvorsky, "Prominent Scientists Sign Declaration that Animals Have Conscious Awareness, Just Like Us," io9.com. http://io9.com/5937356/prominent-scientists-sign-declaration-that -animals-have-conscious-awareness-just-like-us/ (accessed August 23, 2012).

136 *Years before* The Cove *hit theaters:* For more drive hunt footage, please see Hardy Jones's Web site, http://www.bluevoice.org/webfilms.php. Jones has also written a compelling book, *The Voice of the Dolphins,* about his time in

Taiji, and in Futo and Iki Island, two other Japanese dolphin-hunting towns. His site is a trove of information about dolphins and the threats they currently face.

136 *Like O'Barry, Marino was:* Erik Vance, "It's Complicated: The Lives of Dolphins and Scientists," *Discover*, September 2011, 62–76.

141 *"the communal self":* Harry J. Jerison, "The Perceptual Worlds of Dolphins," in Ronald J. Schusterman, Jeanette A. Thomas, and Forrest G. Wood, eds., *Dolphin Cognition and Behavior: A Comparative Approach* (New Jersey: Lawrence Erlbaum Associates, 1986): 160–61.

Jerison's work is absorbing, if pretty heavy sledding. I was particularly taken by his willingness to mix scientific inquiry with philosophical questions. His definition of intelligence, a concept famously tough to pin down, is one example: "The mind and conscious experience are constructions of nervous systems to handle the overwhelming amount of information they possess," Jerison wrote. "Intelligence is a measure of the capacity for such constructions."

143 *"The human brain is the most unsuccessful":* Diane Ackerman, *The Moon by Whalelight* (New York: Vintage Books, 1991), 144.

144 *"My thought was, 'Okay'":* Virginia Morell, "Minds of Their Own," *National Geographic*, March 2008, 36–61.

144 *"An important finding":* Louis M. Herman, "What the Dolphin Knows, or Might Know, in Its Natural World," in Karen Pryor and Kenneth S. Norris, eds., *Dolphin Societies: Discoveries and Puzzles* (Berkeley: University of California Press, 1998), 349–63.

145 *"wide-ranging intellectual":* Louis M. Herman, "What Laboratory Research Has Told Us about Dolphin Cognition," *International Journal of Comparative Psychology* 23 (2010): 310–30.

CHAPTER 7: HIGH FREQUENCY

149 *"Earth, to put the matter succinctly":* Edward O. Wilson, "Beware the Age of Loneliness," *The Economist*, November 18, 2013.

150 *"What, I wondered":* Allan J. Hamilton, "The Sixth Sense," Harvard Medicine, http://hms.harvard.edu/news/harvard-medicine/sixth-sense (accessed April 2014).

150 *"Dolphins play with the energies":* Joan Ocean, *Dolphins into the Future* (Kailua, HI: Dolphin Connection, 1997): 147–48.

153 *"These tones can transform":* Ibid., 123.

155 *she had coauthored a paper:* Lori Marino and Scott O. Lilienfeld, "Dolphin-Assisted Therapy for Autism and Other Developmental Disorders: A Dangerous Fad," *Psychology in Intellectual and Developmental Disabilities* 33, no. 2 (Fall 2007): 2–3.

161 *"the cheetahs of the deep sea":* Natacha Aquilar Soto, Mark P. Johnson, Peter T. Madsen, Francisca Diaz, Ivan Dominguez, Alberto Brito, and Peter Tyack, "Cheetahs of the Deep Sea: Deep Foraging Sprints in Short-Finned

Pilot Whales Off Tenerife," *Journal of Animal Ecology* 77 (September 2008): 936–47.

162   *"I was very interested in Sirius"*: If you dive into the New Age dolphin world, it won't be long before you hear talk of Sirius, the brightest star we can see from Earth. (Sirius is actually a system of two stars, Sirius A and B. A is the brilliant one, massively outshining our sun, albeit from 8.6 light-years away; B is a white dwarf, fainter and smaller.) As the lynchpin of the constellation Canis Major, Sirius is known as the "Dog Star," but it is often rebranded by New Age dolphin lovers. Perhaps this is partly because it is one of the closer stars to us, and theoretically convenient for planet-hopping dolphins, or because of its alleged significance to ancient civilizations (the ancient Egyptians, for instance, based their calendars on it). Most likely, the star's main dolphin affiliation can be traced to the Dogon people of West Africa. The Dogons believe their ancestors were dolphin-like creatures called Nommos, who, as their legends tell it, called Sirius home.

169   *"I tell them"*: The Chumash are joined in their philosophies by other Native American tribes. This passage from Luther Standing Bear, a Sioux (Oglala Lakota) chief born in 1868, is too gorgeous not to share: "Everything was possessed of personality, only differing from us in form. Knowledge was inherent in all things. The world was a library and its books were the stones, leaves, grass, brooks, and the birds and animals that shared, alike with us, the storms and blessings of earth. We learned to do what only the student of nature ever learns, and that was to feel beauty. We never railed at the storms, the furious winds, and the biting frosts and snows. To do so intensified human futility, so whatever came we adjusted ourselves, by more effort and energy if necessary, but without complaint."

   Luther Standing Bear, *Land of the Spotted Eagle* (Lincoln, NE: University of Nebraska Press, 1978), 194–95.

170   *250 decibels:* Decibels are the units used to measure the relative intensity of sound. Explaining how they work requires some fierce math, but in essence the decibel scale is logarithmic, which means that as you progress up it, sounds get exponentially, not incrementally, louder. A 180-decibel sound can burst eardrums. At 250 decibels, underwater air guns approach the threshold of the loudest noise humans are physically capable of making.

170   *"unavoidable adverse impacts"*: Jason Dearen, "Officials Mull Seismic Tests Near California Nuke Plant," Associated Press, October 1, 2012.

170   *The vote was unanimous:* Krista Schwimmer, "California Coastal Commission Silences Pacific Gas & Electric's Airguns," *The Free Venice Beachhead*, December 2012.

CHAPTER 8: THE WORLD'S END

175   *"It is far better"*: Eddie Osifelo, "Dolphin Debate Heats Up," *Solomon Star*, January 20, 2013.

175   *"a killing spree":* Ednal R. Palmer, "300 More Dolphins Slaughtered," *Solomon Star,* January 25, 2013.

175   *Video of the hunts: The Dolphin Hunters,* produced by Drew Ambrose (Al Jazeera English, 101 East, 2014).

176   The Guardian *also weighed:* Suzanne Goldberg, "Solomon Islands Villagers Kill 900 Dolphins in Conservation Dispute," *The Guardian,* January 24, 2013. "International Outrage Over Solomon Islands Mass Dolphin Slaughter," ABC Radio Australia, January 23, 2013.

177   *The more teeth a family displays:* For a concise history of the Solomon Island dolphin hunts, and a detailed account of the February and March 2013 hunts this chapter refers to, please see Marc Oremus, John Leqata, and Scott C. Baker, "Resumption of Traditional Drive Hunting of Dolphins in the Solomon Islands in 2013," *Royal Society Open Science* 2 (May 2015), http://rsos.royal societypublishing.org/content/royopensci/2/5/140524.full.pdf.

178   *In July 2003:* Richard C. Paddock and Richard Boudreaux, "Any Way to Treat a Dolphin?" *Los Angeles Times,* October 17, 2003; *The Dolphin Trade,* television show produced by Joe Rhee and Kimberly Launier (Primetime Live, ABC Entertainment Group, October 27, 2005).

178   *"It's big":* "Solomon Islands to Export 30 Dolphins to Dubai," Yahoo News, October 12, 2007.

178   *O'Barry viewed video:* William Rossiter, "Greed, Corruption and Captivity," *Cetacean Society International* 12, no. 4 (October 2003).

179   *"Like the rest of the world": Blood Dolphins: The Solomons Mission,* directed by Lincoln O'Barry, produced by Raymond Bridgers, Dave Harding, Erik Nelson, Lincoln O'Barry, and Dieu Pham (Animal Planet, September 2010).

179   *"I love animals": The Dolphin Dealer,* directed, produced, and written by Brad Quenville (Vancouver: Omni Film Productions, 2008).

183   *"trying to disrupt":* Associated Press, "Dubai's Dolphin Import Angers Activists," *USA Today,* October 18, 2007.

185   *Online, I had watched a promo video:* "Solomon Islands Dolphins Paradise," YouTube video, 2:05, posted by Christopher Palmer, March 6, 2008, https://www.youtube.com/watch?v=asCKC4yW9vk (last accessed April 2015).

188   *A Canadian documentary crew: The Dolphin Dealer.*

189   *"It's a shame":* Ibid.

190   *the memory still scalded:* There is no indication that Porter and Satu were involved in Makili's attack.

192   *"We will slaughter": Blood Dolphins,* episode 103: "Solomon Islands," part 2, Animal Planet, September 7, 2010.

200   *"an amazing project":* "Honiara's New Kokonut Café," *Solomon Star News,* March 26, 2012.

CHAPTER 9: GREETINGS FROM HAWAII:
WE'RE HAVING A BLAST!

213 *"Race On to Find Gulf Coast"*: Ed Lavandera and Jason Morris, "Race On to Find Gulf Coast Dolphin Killers," CNN, November 28, 2012. http://www .cnn.com/2012/11/28/us/gulf-coast-dolphin-killings/.

213 *"What's Behind Spike"*: Rena Silverman, "What's Behind Spike in Gulf Coast Dolphin Attacks?" National Geographic News, March 29, 2013. http://news .nationalgeographic.com/news/2013/03/130329-dolphin-attacks-gulf-coast -marine-mammals-oceans-science/.

213 *Similar dolphin-loathing sentiments:* Luiz Cláudio Pinto de Sá Alves, Camilah Antunes Zappes, and Artur Andriolo, "Conflicts Between River Dolphins (Cetacea: Odontoceti) and Fisheries in the Central Amazon: A Path Toward Tragedy?" *Zoologica* 29 (October 2012): 420–29.

214 *One startling effect:* Ruth H. Carmichael, William M. Graham, Allen Aven, Graham Worthy, Stephan Howden, "Were Multiple Stressors a 'Perfect Storm' for Northern Gulf of Mexico Bottlenose Dolphins (Tursiops truncatus) in 2011?" Plos One, Published July 18, 2012. http://journals.plos.org/plosone/ article?id=10.1371/journal.pone.0041155.

214 *For the past five years:* "Cetacean Unusual Mortality Event in Northern Gulf of Mexico (2010–Present)," NOAA Fisheries online. http://www.nmfs.noaa .gov/pr/health/mmume/cetacean_gulfofmexico.htm.

215 *"I've never seen"*: Lori H. Schwacke et al., "Health of Common Bottlenose Dolphins (Tursiops truncatus) in Barataria Bay, Louisiana, Following the Deepwater Horizon Oil Spill," *Environmental Science and Technology* 48 (2014): 93–103.

217 *Perhaps the Navy could build:* Unfortunately, this is not a hypothetical example. In 2010, after the Navy announced plans to build its Undersea Warfare Training Range fifty miles offshore from Jacksonville, Florida, right next to the whales' only known calving grounds, a host of environmental groups sued it—and lost. The training range was allowed to proceed. North Atlantic right whales are a critically endangered species, with only 300 individuals remaining.

217 *By the Navy's own estimates:* Please note that these numbers reflect only one proposal, for one area of the Pacific Ocean (Southern California and Hawaii). The U.S. Navy has also proposed similar plans for the Atlantic, the Northwest Pacific, and other areas. Add to that the navies from other countries, each conducting sonar and missile testing programs of its own.

217 *As for underwater bomb:* I read about this response in *War of the Whales,* a richly detailed narrative about the use of military sonar in the ocean, its impact on cetaceans, and the concerted efforts of scientists and environmental groups to hold the Navy to account. Joshua Horowitz, *War of the Whales* (New York: Simon & Schuster, 2014), 123.

226 *"To say that dolphins"*: Patrick W. B. Moore, "Dolphin Psychophysics: Concepts for the Study of Dolphin Echolocation," *Dolphin Societies* (Berkeley: University of California Press, 1998), 365.

228 *"swimmer nullification"*: "Dolphins Aweigh," produced by William Brown, *60 Minutes*, February 18, 1973. (CBS)

228 *"Marine mammals are actually"*: U.S. Navy Marine Mammal Program, FAQ page, http://www.public.navy.mil/spawar/Pacific/71500/Pages/faqs.aspx.

230 *Scientists refer to this:* For an excellent, comprehensive overview of this problem, see: Michael Jasny, with Joel Reynolds, Cara Horowitz, and Andrew Wetzler, *Sounding the Depths II: The Rising Toll of Sonar, Shipping and Industrial Ocean Noise on Marine Life*, National Resources Defense Council, November 2005. Available for download at: www.nrdc.org/wildlife/marine/sound/sound.pdf.

CHAPTER 10: CHANGE OF HEART

236 *at least sixteen orcas:* Though Marineland is often unwilling to confirm or deny the deaths of its marine mammals, an advocacy Web site, Marineland in Depth, has compiled an archive of news clippings about the park, dating back to 1963: www.marinelandindepth.com/2013/08/the-marineland-news-archive.html. Regarding orca deaths, I arrived at the conservative number of sixteen by comparing these archived news clippings with lists maintained by various organizations that have kept track of Marineland's orcas over the years. These sources are Ceta-Base, Zoocheck Canada, and the Orca Project Database. (Wikipedia also contains a list of Marineland's deceased orcas, although it is incomplete.) Additionally, these sites report three orca stillbirths at the park. Several orcas have never been accounted for, and are designated "missing/presumed dead."

On occasions when Marineland has acknowledged orca deaths, its longtime spokeswoman, Ann Marie Rondinelli, has expressed the park's sadness. After three-year-old Malik died, for instance, Rondinelli told the press: "We all work very closely with these animals. To us, they're very special, they're almost like a family member . . . You have people crying and quite upset." Christine Cox, "Marineland Says Orca Whale Was 'Like Family,'" *The Spectator*, March 11, 2000.

236 *"You have to understand"*: Linda Diebel and Liam Casey, "Marineland: Inside the Controversy," *Toronto Star*, Star Dispatches e-book, published 2013.

237 *numerous whistleblowers:* In the *Toronto Star* series, Diebel and Casey state that by the end of 2012, fifteen whistleblowers had spoken to them about Marineland.

237 *When Holer refused:* Alison Langley, "Marineland Told to Hand Over Whale to U.S.," *St. Catharines Standard*, September 29, 2011. In affidavits related to the court case, SeaWorld executives expressed concerns about Ikaika's elevated white blood cell count, his mental health, and Marineland's facilities. The whale's mood was no minor matter, given that Ikaika apparently had a history of aggression—and his father was Tilikum. Liam Casey, "Custody of Killer Whale Plays Out in Court," *Toronto Star*, July 16, 2011, and Liam Casey, "Send Killer Whale Back to Florida, Court Tells Marineland," *Toronto Star*, September 28, 2011.

237 *"Once an animal turns":* Tony Ricciuto, "Bear Death Act of Nature Says Marineland's Owner," *Niagara Falls Review,* September 3, 1993.

238 *more than a thousand carcasses:* Linda Diebel and Liam Casey, "Marineland: Environment Ministry Launches Probe into Mass Animal Graves," *Toronto Star,* December 20, 2012. When Ontario's environment ministry was alerted to the existence of these clandestine burial sites, it launched an immediate inspection, and ordered Marineland to stop burying its animals on the grounds without a permit. "We are concerned about the locations of the sites because [they] are so close to a water course," reported a government spokesperson. Marineland responded: "One of life's sad truths is that animals sometimes die in zoos, just as they do in the wild."

239 *Junior died in 1994:* Sands filmed Junior for a final time in May 1994. "The last time I was there, the last time I saw Junior, he didn't move," she told me in an interview. "At one point, his rostrum was just millimeters from the periphery of the tank, at one point he just rolled over and opened up his mouth and then rolled back and was just floating. Like he was the floating dead. There was no life left in this animal." When Sands returned to the Barn two months later, in July 1994, Junior was gone. Two of the park's trainers told her that the orca had died.

239 *Marineland has never commented:* When asked about Junior by the *Toronto Star,* Marineland's spokeswoman, Ann Marie Rondinelli, deflected the questions, saying: "We are focusing on what we do best—ensuring our guests enjoy their visits to our park, confident in the knowledge that all of our animals are well cared for." Linda Diebel, "Swimming Alone," *Toronto Star,* August 25, 2012.

239 *Over the years:* This tally comes from Zoocheck Canada, an advocacy group that regularly monitors the country's marine parks, sending in scientists and other experts. Though Marineland has never responded to requests to confirm or deny the deaths of several orcas whose status is "missing/presumed dead," Zoocheck breaks the numbers down as follows: "Total orcas ever exhibited: 29. Exported or re-exported: 8. Died/presumed dead at Marineland: 19. Total dead/presumed dead: 26 (90%). Total still alive at Marineland: 2. Total still alive elsewhere: 1." When I visited Niagara Falls in May 2013, a single orca, Kiska, remained, and not two as stated. http://www.zoocheck.com/campaigns_whaleswild_mlinventory.pdf.

240 *Diebel and Casey found a video:* Linda Diebel and Liam Casey, "Marineland Whale Bleeding for Months," *Toronto Star,* October 18, 2012. Santos described Kiska's tail as "gushing" blood from wounds that were worsening. The orca, Santos said, repeatedly scratched herself against sharp fiberglass grates in her tank. Video clearly showed blood streaming from Kiska's tail flukes. Marineland refuted the allegation, calling it "seriously inaccurate."

242 *orcas stick close:* On occasion transient orcas do leave their mothers, particularly females with their own offspring.

243 *"the most amazing animals":* Robert Pitman, "The Top, Top Predator," *Journal of the American Cetacean Society* 40, no. 1 (Spring 2011): 5. http://www.orcanetwork.org/Main/PDF/WhalewatcheVol40No12011.pdf.

245 *One recent study:* S. De Guide, A. Lagacé, and P. Béland, "Tumors in St. Law- rence Beluga Whales," *Veterinary Pathology Online* 31 (1994): 444–49.

246 *"Kiska is now quite elderly":* "Marineland Opens Up About Kiska," *Niagara Falls Review,* July 24, 2014.

251 *His words trailed off:* I had never quite grasped the logic behind Porter's Free the Pod campaign, mainly because the easiest way to let dolphins go is simply to remove the nets, and he could have done so at any moment. According to Porter, however, the situation was more complicated than that. If the villagers and the government were not paid off, he claimed, the dolphins would have been promptly recaptured. "They would round up all of them," he told me. "The fishermen aren't gonna let dolphins go that are good for teeth." When I asked him if he was describing a type of extortion (pay us or we'll kill your dolphins), he seemed surprised—but that is exactly what it sounded like to me. Ultimately, after listening to Porter talk extensively about Free the Pod, I still don't understand it, or why releasing the dolphins even in the absence of payola wasn't a better option than having them all die in Gavutu's lagoon.

253 *Lolita's mother:* For more information about the Southern Resident popula- tion, and orca research in the Pacific Northwest, see the Center for Whale Research's Web site: www.whaleresearch.com.

253 *It was possible, even likely:* To read the details of this plan, see: http://savelolita .org/the-plan.

253 *In 2014, the National Aquarium:* Virginia Morell, "Q&A: National Aquarium CEO Discusses Dolphins' Retirement," nationalgeographic.com, May 21, 2014. http://news.nationalgeographic.com/news/2014/05/140520-bottlenose -dolphins-national-aquarium-sanctuary-captivity-oceans-science/ (accessed April 2015).

254 *"No cetacean should be held":* To read the entire declaration, see: www .cetaceanrights.org.

254 *"Dolphins are nonhuman persons":* Associated Press, "Whale and Dolphins Should Have Legal Rights," *The Guardian,* February 20, 2012. http://www .theguardian.com/world/2012/feb/21/whales-dolphins-legal-rights.

CHAPTER 11: THERA

262 *"We think we have understood":* Thomas Berry, *The Sacred Universe* (New York: Columbia University Press, 2009), 171.

268 *"The Minoans were not":* Craig S. Barnes, *In Search of the Lost Feminine: Decod- ing the Myths that Radically Reshaped Civilization* (Golden, CO: Fulcrum Pub- lishing, 2006), 1.

272 *"Large dolphins and numerous":* Arthur Evans, as quoted in Joseph Alexander MacGillivray, *Minotaur: Sir Arthur Evans and the Archaeology of the Minoan Myth* (New York: Farrar, Straus and Giroux, 2000), 217.

SELECTED BIBLIOGRAPHY

BOOKS

Barnes, Craig S. *In Search of the Lost Feminine: Decoding the Myths That Radically Reshaped Civilization* (Golden, CO: Fulcrum Publishing, 2006).

Bearzi, Maddalena, and Craig B. Stanford. *Beautiful Minds: The Parallel Lives of Great Apes and Dolphins* (Cambridge, MA: Harvard University Press, 2008).

Bekoff, Marc; Colin Allen; and Gordon M. Burghardt, eds. *The Cognitive Animal: Empirical and Theoretical Perspectives on Animal Cognition* (Cambridge, MA: MIT Press, 2002).

Berry, Thomas. *The Sacred Universe: Earth, Spirituality, and Religion in the Twenty-first Century* (New York: Columbia University Press, 2009).

Burnett, D. Graham. *The Sounding of the Whale: Science and Cetaceans in the Twentieth Century* (Chicago: University of Chicago Press, 2012).

Caldwell, David K., and Melba C. Caldwell. *The World of the Bottlenosed Dolphin* (Philadelphia: J. B. Lippincott Company, 1972).

Carwardine, Mark. *Whales, Dolphins, and Porpoises* (New York: Doring Kindersley Publishing, 2002).

Castleden, Rodney. *Minoans: Life in Bronze Age Crete* (London: Routledge, 1994).

———. *Atlantis Destroyed* (London: Routledge, 2001).

Chapskii, K. K., and V. E. Sokolov, eds. *Morphology and Ecology of Marine Mammals: Seals, Dolphins, Porpoises* (New York: John Wiley & Sons, 1973).

Devine, Eleanore, and Martha Clark, eds. *The Dolphin Smile: Twenty-nine Centuries of Dolphin Lore* (New York: MacMillan Company, 1967).

Doumas, Chr. G. *The Early History of the Aegean in the Light of Recent Finds from Akrotiri, Thera* (Athens: Society for the Promotion of Studies on Prehistoric Thera, 2008).

Dudzinski, Kathleen M., and Toni Frohoff. *Dolphin Mysteries: Unlocking the Secrets of Communication* (New Haven, CT: Yale University Press, 2008).

Eagleman, David. *Incognito: The Secret Lives of the Brain* (New York: Vintage Books, 2012).

Eiseley, Loren. *The Star Thrower* (New York: Harcourt Brace & Company, 1978).

Ellis, Richard. *Dolphins and Porpoises* (New York: Alfred A. Knopf, 1982).

Fichtelius, Karl-Erik, and Sverre Sjolander. *Smarter Than Man? Intelligence in Whales, Dolphins, and Humans* (New York: Pantheon Books, 1972).

Gregg, Justin. *Are Dolphins Really That Smart? The Mammal Behind the Myth* (Oxford, U.K.: Oxford University Press, 2013).

Hardy, D. A.; C. G. Doumas; J. A. Sakellarakis; and P. M. Warren, eds. *Thera and the Aegean World III: Volume One, Archaeology* (London: Thera Foundation, 1990).

Herman, Louis M., ed. *Cetacean Behavior: Mechanisms & Functions* (Malabar, FL: Robert E. Krieger Publishing Company, 1980).

Herzing, Denise L. *Dolphin Diaries: My Twenty-five Years with Spotted Dolphins in the Bahamas* (New York: St. Martin's Press, 2011).

Horowitz, Joshua. *War of the Whales* (New York: Simon & Schuster, 2014).

Johnson, William. *The Rose-Tinted Menagerie* (London: Heretic Books, 1990).

Jones, Hardy. *The Voice of the Dolphins* (St. Augustine, FL: BlueVoice.org, 2011).

Kirby, David. *Death at SeaWorld: Shamu and the Dark Side of Killer Whales in Captivity* (New York: St. Martin's Press, 2012).

Lilly, John C. *Man and Dolphin* (New York: Doubleday & Company, 1961).

———. *The Mind of the Dolphin: A Nonhuman Intelligence* (New York: Avon Books, 1967).

———. *The Center of the Cyclone: An Autobiography of Inner Space* (New York: Julian Press, 1972).

———. *Lilly on Dolphins: Humans of the Sea* (New York: Anchor Press, 1975).

———. *Communication Between Man and Dolphin: The Possibility of Talking with Other Species* (New York: Julian Press, 1978).

MacGillivray, Joseph Alexander. *Minotaur: Sir Arthur Evans and the Archaeology of the Minoan Myth* (New York: Farrar, Straus and Giroux, 2000).

Mann, Janet; Richard C. Connor; Peter L. Tyack; and Hal Whitehead, eds. *Cetacean Societies: Field Studies of Dolphins and Whales* (Chicago: University of Chicago Press, 2000).

Marinatos, Sp. *Life and Art in Prehistoric Thera* (London: Oxford University Press, 1971).

Messenger, Cheryl, and Terran McGinnis. *Marineland* (Charleston: Arcadia Publishing, 2011).

Norris, Kenneth S. *Dolphin Days: The Life and Times of the Spinner Dolphin* (New York: W. W. Norton & Company, 1991).

Norris, Kenneth S.; Bernd Würsig; Randall S. Wells; and Melany Würsig. *The Hawaiian Spinner Dolphin* (Berkeley: University of California Press, 1994).

O'Barry, Richard, with Keith Coulbourn. *Behind the Dolphin Smile* (Los Angeles: Renaissance Books, 1999).

Ocean, Joan. *Dolphins into the Future* (Hawaii: Dolphin Connection, 1997).

Pryor, Karen, and Kenneth S. Norris, eds. *Dolphin Societies: Discoveries and Puzzles* (Berkeley: University of California Press, 1998).

Reiss, Diana. *The Dolphin in the Mirror: Exploring Dolphin Minds and Saving Dolphin Lives* (Boston: Houghton Mifflin Harcourt, 2011).

Schusterman, Ronald J.; Jeanette A. Thomas; and Forrest G. Wood, eds. *Dolphin Cognition and Behavior: A Comparative Approach* (Hillsdale, NJ: Lawrence Erlbaum Associates, 1986).

Scully, Matthew. *Dominion: The Power of Man, the Suffering of Animals, and the Call to Mercy* (New York: St. Martin's Press, 2002).

Smokler, Rachel. *To Touch a Wild Dolphin: A Journey of Discovery with the Sea's Most Intelligent Creatures* (New York: Anchor Books, 2002).

Vougioukalakis, George A. *The Minoan Eruption of the Thera Volcano and the Aegean World* (Athens: Society for the Promotion of Studies on Prehistoric Thera, 2008).

Weyler, Rex. *Song of the Whale: The Dramatic Story of Dr. Paul Spong* . . . (New York: Anchor Press, 1986).

White, Thomas I. *In Defense of Dolphins: The New Moral Frontier* (Malden, MA: Blackwell Publishing, 2007).

Whitehead, Hal, and Luke Rendell. *The Cultural Lives of Whales and Dolphins* (Chicago: University of Chicago Press, 2015).

William Andrews Clark Memorial Library. *The Dolphin in History: Papers Delivered by Ashley Montagu and John C. Lilly at a Symposium at the Clark Library, 13 October, 1962.*

VIDEOS

Blackfish: Directed by Gabriela Cowperthwaite, produced by Manuel Oteyza and Gabriela Cowperthwaite (CNN Films: July 2013). www.blackfishmovie.com.

Blood Dolphins: Saving the Solomons, and *The Solomons Mission*: Directed by Lincoln O'Barry, produced by Raymond Bridgers, Dave Harding, Erik Nelson, Lincoln O'Barry, and Dieu Pham (Animal Planet: September 2010). www.animalplanet .com/tv-shows/blood-dolphins.

The Cove: Directed by Louis Psihoyos, produced by Paula DuPré Pesman and Fisher Stevens (Participant Media: July 2009). www.thecovemovie.com.

The Dolphin Dealer: Directed and produced by Brad Quenville (Omni Film Productions, 2008).

ORGANIZATIONS

BLUE FRONTIER FOUNDATION: www.bluefront.org

BLUEVOICE: www.bluevoice.org

CASCADIA RESEARCH COLLECTIVE: www.cascadiaresearch.org

CENTER FOR WHALE RESEARCH: www.whaleresearch.com

EARTH ISLAND INSTITUTE: www.earthisland.org/immp

EARTHJUSTICE: www.earthjustice.org

HUMANE SOCIETY OF THE UNITED STATES: www.humanesociety.org

KIMMELA CENTER FOR ANIMAL ADVOCACY: www.kimmela.org

NATURAL RESOURCES DEFENSE COUNCIL: www.nrdc.org

NONHUMAN RIGHTS PROJECT: www.nonhumanrightsproject.org

OCEANA: www.oceana.org

ORCALAB: www.orcalab.org

RIC O'BARRY'S DOLPHIN PROJECT: www.dolphinproject.com

SURFERS FOR CETACEANS: www.s4cglobal.org

SURFRIDER: www.surfrider.org

WATERKEEPER ALLIANCE: www.waterkeeper.org

WHALE TRUST MAUI: www.whaletrust.org

WILD DOLPHIN PROJECT: www.wilddolphinproject.com

WISHTOYO: www.wishtoyo.org

PHOTOGRAPHY CREDITS

INSERT

Page 1   *Top and bottom:* Rennio Maifredi
Page 2   *Top:* The Dolphinproject.com *Bottom:* Dianette Wells
Page 3   *Top:* © Flip Schulke. Photographed by Flip Schulke for *Life* magazine, 1961
          Bottom: Courtesy of the State Archives of Florida
Page 4   Courtesy of the State Archives of Florida
Page 5   Courtesy of the State Archives of Florida
Page 6   *Top:* Jeannine Masset and Rudi Schamhart *Bottom:* Jeannine Masset and Rudi
          Schamhart
Page 7   *Top:* Courtesy of the State Archives of Florida *Bottom:* The Dolphinproject.com
Page 8   *Top, center, and bottom:* The Dolphinproject.com
Page 9   *Top:* The Dolphinproject.com *Center, left:* The Dolphinproject.com *Center, right:*
          Susan Casey *Bottom:* The Dolphinproject.com
Page 10   *Top, left:* Courtesy of Lori Marino *Top, right:* Courtesy of Lori Marino *Bottom:*
          Courtesy of the State Archives of Florida
Page 11   *Top:* Anthony Turner *Bottom:* The Dolphinproject.com
Page 12   *Top:* Susan Casey *Bottom, left:* Anthony Turner *Bottom, right:* The
          Dolphinproject.com
Page 13   *Top:* Robin Baird *Center, left:* Daniel Webster *Center, right:* Annie M. Gorgone
          Bottom: Robin Baird
Page 14   *Top and center:* Susan Casey *Bottom:* Rennio Maifredi
Page 15   *Top, center, and bottom:* Rennio Maifredi
Page 16   Robin Baird

ABOUT THE AUTHOR

Susan Casey, author of the *New York Times* bestsellers *The Wave: In Pursuit of the Rogues, Freaks, and Giants of the Ocean* and *The Devil's Teeth: A True Story of Obsession and Survival Among America's Great White Sharks*, is the former editor in chief of *O, The Oprah Magazine*. She is a National Magazine Award–winning journalist whose work has been featured in the *Best American Science and Nature Writing*, *Best American Sports Writing*, and *Best American Magazine Writing* anthologies; and has appeared in *Esquire, Sports Illustrated, Fortune,* and *National Geographic*.